Molecular Biocatalysis 2.0

Molecular Biocatalysis 2.0

Editors

Vladimír Křen
Pavla Bojarová

MDPI • Basel • Beijing • Wuhan • Barcelona • Belgrade • Manchester • Tokyo • Cluj • Tianjin

MDPI

Editors
Vladimír Křen
National Centre of Biocatalysis and Biotransformation
Czech Republic

Pavla Bojarová
National Centre of Biocatalysis and Biotransformation
Czech Republic

Editorial Office
MDPI
St. Alban-Anlage 66
4052 Basel, Switzerland

This is a reprint of articles from the Special Issue published online in the open access journal *International Journal of Molecular Sciences* (ISSN 1422-0067) (available at: https://www.mdpi.com/journal/ijms/special_issues/molecularbiocatalysis).

For citation purposes, cite each article independently as indicated on the article page online and as indicated below:

LastName, A.A.; LastName, B.B.; LastName, C.C. Article Title. *Journal Name* **Year**, *Article Number*, Page Range.

ISBN 978-3-03943-278-3 (Hbk)
ISBN 978-3-03943-279-0 (PDF)

Contents

About the Editors

Vladimír Křen is the head of the Laboratory of Biotransformation, Institute of Microbiology, Czech Academy of Sciences, Prague, and a full Professor of Medicinal Chemistry at Palacky University, Olomouc. He spent his postdoctorals and sabbaticals at Hamburg University, Warwick University, University of Bologna, Université Paris-Sud Orsay, and Okayama University, Japan. His main interests include biocatalysis and biotransformation; immobilized microbial cells, their use in the production and biotransformation of natural products; biotransformation of natural products by enzymes and microorganisms; preparation of glycosidases of microbial origin and their use in the glycosylation of natural compounds: glycoconjugates, multivalent compounds, ergot alkaloids, flavonoids, antioxidants and chemoprotectants. He is the (co-)author of ca. 400 publications that have been cited over 8500 times (h-index 43).

Pavla Bojarová graduated at Charles University in Prague, Faculty of Sciences, in 2002, and obtained her PhD in biochemistry in 2006. For her postdoctoral study, she joined Prof. Spencer Williams' group at the University of Melbourne, to work on the time-dependent inactivation of sulfatases. Since her undergraduate years, she has been working in the Laboratory of Biotransformation with Prof. Vladimír Křen, Academy of Sciences of the Czech Republic, Prague. Her main research interests include the (chemo-)enzymatic synthesis of complex glycostructures and glyco-nanomaterials; analysis and mutagenesis of carbohydrate-active enzymes and the application of bioactive glycoconjugates in the study of lectin–carbohydrate interactions. She has (co-)authored 44 papers and four book chapters, and been cited over 1100 times (h-index 20).

International Journal of
Molecular Sciences

MDPI

Editorial

Biocatalysis: "A Jack of all Trades..."

Vladimír Křen * and Pavla Bojarová

Laboratory of Biotransformation, Institute of Microbiology of the Czech Academy of Sciences, Vídenská 1083, 14220 Prague, Czech Republic; bojarova@biomed.cas.cz

* Correspondence: kren@biomed.cas.cz

Received: 13 July 2020; Accepted: 15 July 2020; Published: 20 July 2020

Biotransformation has accompanied mankind since the Neolithic community, when people settled down and began to engage in agriculture. Before that time, "biotransformation" was just a "piece of rotten meat". Alcohol, accompanying humanity for more than nine millennia, may have been the two-edged sword that helped to drive the development of fermentation technologies, ... followed by leavened bread ... and then vinegar.

Modern biocatalysis started with "Enzymes" in the second half of 19th century; the term "enzyme" was coined by prof. Kühne [1]. Already in 1858, Pasteur described the first chiral separation of tartaric acid by biotransformation with *Penicillium glaucum* [2]. One of the first patented biotransformation processes—in modern terms—was the production of Neuberg ketol ((*R*)-1-phenyl-l-hydroxy-2-propanone), a precursor of (-)-ephedrin, which is still produced by the same technology [3]. The fulminant development of biocatalytic procedures continued with the industrial production of ascorbic acid (the biotransformation of D-sorbitol into L-sorbose using *Acetobacter suboxydans*); later in the 1950s, biocatalysis started to flourish by steroid-selective bio-oxidation (prednisolon). Then, biotransformations became an indispensable part of our lives, similar to computers and other hi-tech products.

Recent development demonstrates that biotransformation or biocatalysis became a true "Jack of all Trades" as the product portfolio covers virtually all fields of human activities and needs. The Special Issue "Molecular Biocatalysis 2.0" aims to cover a broad array of biocatalytic procedures to demonstrate its versatility and applicability.

A number of papers are devoted to biocatalysis in glycobiology. This is quite logical, since the chemical synthesis of glycostructures still involves demanding multistep cascades of reactions, protection/deprotection steps, and purifications. In contrast, enzymatic systems often provide a single-step selective process, yielding directly a "deprotected" product. This has been nicely demonstrated in the one-pot enzyme cascade synthesis of high molecular weight hyaluronic acid, which is largely used in cosmetic applications [4]. This elegant procedure, an alternative to the use of animal tissues, combines two enzyme modules that in situ generate the respective precursors UDP–GlcA and UDP–GlcNAc using the hyaluronan synthase from *Pasteurella multocida* (*Pm*HAS). This method was optimized to meet the kinetic requirements of *Pm*HAS for high hyaluronic acid productivity and molecular weight. Another glyco-biotechnology applicable in the food industry describes the anchoring of a bacterial chitosanase on the cell walls of *Lactobacillus plantarum*, which is a food-grade expression system [5]. The chitosanase displayed on *L. plantarum* cells is catalytically active and can convert chitosan into chito oligosaccharides—typically chitobiose and chitotriose as the main products—widely used in pharmacology and medicine. The research groups from Madrid and Prague joined their efforts in a detailed mapping of the acceptor specificity of the transglycosylation reaction catalyzed by β-*N*-acetylhexosaminidase from *Talaromyces flavus*—both as wild type and transglycosidase mutant variants [6]. This study, combined with molecular modelling, deciphered a rather enigmatic problem of acceptor regioselectivity in the transglycosylation reactions and, moreover, provided the first example of the enzymatic glycosylation of an *N*-acetylmuramic acid derivative.

The application of biocatalysis in the modification of natural products for the nutraceutics industry was demonstrated in the bioproduction of quercetin and rutinose catalyzed by rutinosidase from *A. niger* [7]. This study also brought a novel concept of "Solid State Biocatalysis", where both the substrate (rutin) and the product (quercetin) remained in suspension, allowing thus working with concentrations of up to 300 g/L (ca 0.5 M). These results demonstrated for the first time the efficiency of the "Solid-State-Catalysis" concept, which is applicable virtually to any biotransformation involving substrates and products of low water solubility.

Another increasingly attractive research area comprises selective enzymatic redox reactions. Both biooxidations and bioreductions generate new interesting substances, which can hardly be obtained by standard chemical methods. Indigo, a dye used, e.g., for a typical blue color of jeans, is currently produced by a century-old petrochemical-based process. Fraaije et al. [8] showed that the bacterial flavin-monooxygenase from *Methylophaga* sp. can be adapted to improve its ability to convert indole (a commodity chemical) into indigo. This study, entangling computational and structure-inspired enzyme redesign improvement, resulted not only in an upgraded biocatalyst but also provided a better understanding of the structural elements and the detailed mechanism of this important monooxygenase. Oxidases working with inorganic substrates form another interesting facet of biocatalysis. The application of the manganese (Mn^{2+})-oxidizing bacteria *Pseudomonas putida* MnB1 as a whole-cell biocatalyst enabled the effective oxidation of β-keto ester with biogenic MnO_2, generated in situ, in high yields [9]. On top of that, cells of *P. putida* MnB1 remain alive and are capable of the continuous catalysis of the β-keto ester forming reaction for several cycles. Horseradish peroxidase (HRP) is an important heme-containing oxidase that has been studied for more than a century [10], and it has a vast applicability both in biochemical (ELISA assays) and biotechnological processes. A research group from Vienna produced recombinant HRP in *E. coli* as a fusion protein with quadruple mutations at the glycosylation sites. This construct showed a twice better thermostability and an eight-fold increased catalytic activity with 2,2′-azino-*bis*(3-ethylbenzothiazoline-6-sulphonic acid) as the reducing substrate when compared to the non-mutated recombinant HRP benchmark enzyme [11]. Oxidases can also be used as highly selective sensors. This is the case of glucose oxidase (GO), which is used in electrochemical glucose sensors, e.g., for monitoring and accurate glycemic control for diabetic patient care. Engineered GO was able to catalyze direct single-step modification with a redox mediator (phenazine ethosulfate) on its surface via a lysine residue rationally introduced into the enzyme [12]. This modified GO showed a quasi-direct electron transfer response, which enables its use in the third-generation sensors. It is considered as the ideal solution since the measurements can be performed in the absence of a free redox mediator.

Nitrilases are crucial enzymes for nitrile metabolism in plants and microorganisms. These enzymes have already found broad application in industry—e.g., in the large-scale production of acrylamide from acrylonitrile and in the production of numerous pharmaceuticals [13]. Nitrilases were for the first time described in *Basidiomycota*, and over 200 putative nitrilases were found in this division via GenBank. The representatives of clade 1 and 2 (Nit*Tv*1 from *Trametes versicolor* and Nit*Ag* from *Armillaria gallica*, respectively) and a putative CynH (Nit*Sh* from *Stereum hirsutum*) were overproduced in *E. coli*, and their substrate specificities were analyzed in detail [14]. This study substantially broadens the repertoire of nitrilases available for biocatalytic applications.

The above examples clearly demonstrate that molecular biocatalysis is a pluripotent methodology, which strongly contributes with its inherent green concept to the sustainability of our daily lives. Its nickname "A Jack of all Trades" is definitely not an overstatement.

Funding: The work in the authors' laboratory is funded by the projects of Ministry of Education, Youth and Sports of the Czech Republic, grants No. LTC18041 and LTC 20069, and by networking COST projects CA16225 and CA18132.

Conflicts of Interest: The authors declare no conflict of interest.

References

1. Kühne, W.F. Ueber das Verhalten verschiedener organisirter und sog. ungeformter Fermente. Ueber das Trypsin (Enzym des Pankreas). In *Verhandlungen des Naturhistorisch-Medicinischen Vereins zu Heidelberg*; Verlag von Carl Winter's Universitätbuchshandlung: Heidelberg, Germany, 1876; No. 8.1.3.; pp. 190–193.
2. Pasteur, L. Mémoire sur la fermentation de l'acide tartrique. *C. R. Séances Acad. Sci.* **1858**, *46*, 615–618.
3. Neuberg, C.; Hirsch, J. Über ein Kohlenstoffketten knüpfendes Ferment (Carboligase). (Knoll, A.G. Chemische Fabriken in Ludvigshafen. German Pat. No. 548459, 9 April 1930). *Biochem. Z.* **1921**, *115*, 282–310.
4. Gottschalk, J.; Zaun, H.; Eisele, A.; Kuballa, J.; Elling, L. Key factors for a one-pot enzyme cascade synthesis of high molecular weight hyaluronic acid. *Int. J. Mol. Sci.* **2019**, *20*, 5664. [CrossRef] [PubMed]
5. Pham, M.L.; Tran, A.M.; Mathiesen, G.; Nguyen, H.M.; Nguyen, T.H. Cell wall anchoring of a bacterial chitosanase in Lactobacillus plantarum using a food-grade expression system and two versions of an LP × TG anchor. *Int. J. Mol. Sci.* **2020**, *21*, 3773. [CrossRef]
6. Garcia-Oliva, C.; Hoyos, P.; Petrásková, L.; Kulik, N.; Pelantová, H.; Cabanillas, A.H.; Rumbero, Á.; Křen, V.; Hernáiz, M.J.; Bojarová, P. Acceptor specificity of β-*N*-acetylhexosaminidase from *Talaromyces flavus*: A rational explanation. *Int. J. Mol. Sci.* **2019**, *20*, 6181. [CrossRef] [PubMed]
7. Kapešová, J.; Petrásková, L.; Markošová, K.; Rebroš, M.; Kotík, M.; Bojarová, P.; Křen, V. Biocatalytic production of quercetin with rutinosidase: Novel concept of "solid state biocatalysis". *Int. J. Mol. Sci.* **2019**, *20*, 1112. [CrossRef] [PubMed]
8. Lončar, N.; van Beek, H.L.; Fraaije, M.W. Structure-based redesign of a self-sufficient flavin-containing monooxygenase towards indigo production. *Int. J. Mol. Sci.* **2019**, *20*, 6148. [CrossRef] [PubMed]
9. Guo, J.; Guo, H.; Liu, J.; Zhong, F.; Wu, Y. Manganese(II) oxidizing bacteria as whole-cell catalyst for β-keto ester oxidation. *Int. J. Mol. Sci.* **2020**, *21*, 1709. [CrossRef] [PubMed]
10. Bach, A.; Chodat, R. Untersuchungen über die Rolle der Peroxyde in der Chemie der lebenden Zelle. IV. Ueber Peroxydase. *Ber. Deutsch. Chem. Gesell.* **1903**, *36*, 600–605. [CrossRef]
11. Humer, D.; Spadiut, O. Improving the performance of horseradish peroxidase by site-directed mutagenesis. *Int. J. Mol. Sci.* **2019**, *20*, 916. [CrossRef] [PubMed]
12. Suzuki, N.; Lee, J.; Loew, N.; Takahashi-Inose, Y.; Okuda-Shimazaki, Y.; Kojima, K.; Mori, K.; Tsugawa, W.; Sode, K. Engineered glucose oxidase capable of quasi-direct electron transfer after a quick-and-easy modification with a mediator. *Int. J. Mol. Sci.* **2020**, *21*, 1137. [CrossRef] [PubMed]
13. Martínková, L.; Křen, V. Biotransformations with nitrilases. *Curr. Opin. Chem. Biol.* **2010**, *14*, 130–137. [CrossRef] [PubMed]
14. Rucká, L.; Chmátal, M.; Kulik, N.; Petrásková, L.; Pelantová, H.; Novotný, P.; Příhodová, R.; Pátek, M.; Martínková, L. Genetic and functional diversity of nitrilases in Agaricomycotina. *Int. J. Mol. Sci.* **2019**, *20*, 5990. [CrossRef] [PubMed]

International Journal of
Molecular Sciences

MDPI

Article

Key Factors for a One-Pot Enzyme Cascade Synthesis of High Molecular Weight Hyaluronic Acid

Johannes Gottschalk [1], Henning Zaun [2], Anna Eisele [1], Jürgen Kuballa [2] and Lothar Elling [1,*]

1 Laboratory for Biomaterials, Institute of Biotechnology and Helmholtz-Institute for Biomedical Engineering, RWTH Aachen University, Pauwelsstraße 20, 52074 Aachen, Germany; j.gottschalk@biotec.rwth-aachen.de (J.G.); a.eisele@biotec.rwth-aachen.de (A.E.)

2 Research and Development Department, GALAB Laboratories GmbH, Am Schleusengraben 7, 21029 Hamburg, Germany; henning.zaun@galab.de (H.Z.); juergen.kuballa@galab.de (J.K.)

* Correspondence: l.elling@biotec.rwth-aachen.de; Tel.: +49-241-80-28350

Received: 24 October 2019; Accepted: 11 November 2019; Published: 12 November 2019

Abstract: In the last decades, interest in medical or cosmetic applications of hyaluronic acid (HA) has increased. Size and dispersity are key characteristics of biological function. In contrast to extraction from animal tissue or bacterial fermentation, enzymatic in vitro synthesis is the choice to produce defined HA. Here we present a one-pot enzyme cascade with six enzymes for the synthesis of HA from the cheap monosaccharides glucuronic acid (GlcA) and *N*-acetylglucosamine (GlcNAc). The combination of two enzyme modules, providing the precursors UDP–GlcA and UDP–GlcNAc, respectively, with hyaluronan synthase from *Pasteurella multocida* (PmHAS), was optimized to meet the kinetic requirements of PmHAS for high HA productivity and molecular weight. The Mg^{2+} concentration and the pH value were found as key factors. The HA product can be tailored by different conditions: 25 mM Mg^{2+} and 2-[4-(2-hydroxyethyl)piperazin-1-yl]ethanesulfonic acid (HEPES)-NaOH pH 8 result into an HA product with high M_w HA (1.55 MDa) and low dispersity (1.05). Whereas with 15 mM Mg^{2+} and HEPES–NaOH pH 8.5, we reached the highest HA concentration (2.7 g/L) with a yield of 86.3%. Our comprehensive data set lays the basis for larger scale enzymatic HA synthesis.

Keywords: hyaluronic acid; in vitro synthesis; one-pot multi-enzyme; optimization; enzyme cascade

1. Introduction and Motivation

Hyaluronic acid (HA) is a natural, non-sulfated, linear polymer consisting of repeating disaccharide units of [-3)GlcNAc(β1-4)GlcA(β1-]$_n$. In contrast to other glycosaminoglycans, HA is a polysaccharide with a size of up to 10^8 Da. Due to the anionic character, HA is able to bind large amounts of water, resulting in a viscoelastic gel [1–3]. HA occurs in many species. In humans, HA forms are found in the extracellular matrix of connective tissues [3]. Some pathogenic bacteria (*Pasteurella multocida* and *Streptococcus* strains) camouflage themselves with HA capsules to evade the host´s immune reaction [4]. Because of the unique rheological behavior and the non-immunogenic feature, HA is widely used for medical and cosmetic applications such as drug/cosmetic agents, ophthalmic surgery, and tissue engineering [5–10]. With the increasing demand for HA, the market is expected to develop from USD 7.2 billion in 2016 to USD 15.5 billion in 2025 [11].

Current industrial production of high molecular weight HA is based on harsh extraction from rooster combs or bacterial fermentation with *Streptococcus*. Contaminations with bird proteins and exotoxins can cause allergic reactions or infections. These processes result in a highly dispersed HA product, which affects the biological properties of HA. Sometimes long and short HA chains have even counteractive effects, making highly disperse products less predictable for medical uses [12–15]. Fermentation processes with genetically modified microbial strains aiming for high HA titer and low dispersity are under development [12,16].

Enzymatic in vitro synthesis of HA with low dispersity was accomplished with a soluble class 2 hyaluronan synthase (PmHAS) from *P. multocida* [15]. PmHAS binds the nucleotide sugar substrates UDP–GlcA and UDP–GlcNAc at two active sites, respectively, for HA polymerization [17–22]. However, HA in vitro production is limited by the availability and high consumption of the expensive substrates UDP–GlcA and UDP–GlcNAc. In this respect, in situ generations of both nucleotide sugars were coupled to PmHAS in a one-pot synthesis of HA reaching a molecular weight between 0.02 and 0.5 MDa with 70% yield (1.4 g/L) [23]. For UDP–GlcA production, glucuronic acid kinase (AtGlcAK) and UDP–sugar pyrophosphorylase (AtUSP) from *Arabidopsis thaliana*, as well as pyrophosphatase (PmPpA) from *P. multocida* are proven candidates [24–30]. For UDP–GlcNAc, GlcNAc-1-phosphate kinase (BlNahK) from *Bifidobacterium longum* and UDP–GlcNAc pyrophosphorylases from *Streptococcus zooepidemicus* (SzGlmU) or *Campylobacter jejuni* (CjGlmU) as well as PmPpA were successfully applied [23,31–34]. We recently demonstrated the in vitro one-pot synthesis of HA from sucrose and GlcNAc with in situ regeneration of UDP–GlcA. We obtained HA with a molecular weight of 2 MDa with a low dispersity (1.02) and HA titer of 4 g/L after 8 h. We showed that substrate inhibition by UDP–GlcA and a high K_m value for UDP–GlcNAc afford a favorable UDP–sugar ratio for the production of HA by PmHAS. In addition, in the presence of Mg^{2+} or Mn^{2+}, K^+ was found to enhance the polymerization rate of PmHAS [31].

In order to circumvent the cofactor regeneration of NAD^+ for UDP–GlcA synthesis, we aimed in the present work for an optimized one-pot synthesis of HA with six recombinant enzymes (Scheme 1). The one-pot synthesis contains three enzyme modules (EM): UDP–GlcA-, UDP–GlcNAc-, and hyaluronan-module. In the EM UDP–GlcA and EM UDP–GlcNAc, AtGlcAK and BlNahK phosphorylate the monosaccharides GlcA and GlcNAc, respectively, to the corresponding sugar-1-phosphates with ATP consumption. AtUSP and SzGlmU convert them to UDP–GlcA and UDP–GlcNAc, respectively, using uridine triphosphate (UTP). PmPpA hydrolyzes the AtUSP and SzGlmU inhibiting byproduct pyrophosphate (PP_i). Finally, PmHAS utilizes the nucleotide sugars for HA chain polymerization. As a straightforward development, we aimed to optimize and control this enzyme cascade for high yields, high HA concentration and high molecular weight of HA. To investigate multiple parameters for optimization of each enzyme module (EM) we analyzed enzyme reactions by high-throughput multiplexed capillary electrophoresis (MP-CE) monitoring nucleotides and nucleotide sugars [35].

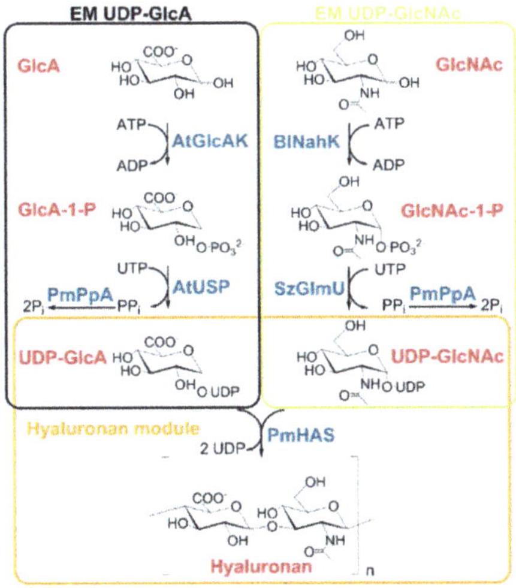

Scheme 1. Enzyme cascade for one-pot hyaluronic acid synthesis. The one-pot synthesis is accomplished by three enzyme modules (EM): UDP–GlcA, UDP–GlcNAc, and hyaluronan module. The EM UDP–GlcA contains the enzymes AtGlcAK, AtUSP (both *Arabidopsis thaliana*), and PmPpA (*Pasteurella multocida*). The EM UDP–GlcNAc contains the enzymes BlNahK (*Bifidobacterium longum*), SzGlmU (*Streptococcus zooepidemicus*), and PmPpA. In the hyaluronan module PmHAS (*P. multocida*) polymerizes the HA chain using UDP–GlcA and UDP–GlcNAc as donor substrates.

2. Results and Discussion

2.1. Enzyme Production

The recombinant enzymes were produced in *E. coli* BL21 (DE3) and purified by immobilized metal affinity chromatography (IMAC). The results of the sodium dodecyl sulfate polyacrylamide gel electrophoresis (SDS-PAGE) and Western blot analyses are summarized in the supplemental data (Figure S1).

2.2. Characterization of AtGlcAK, AtUSP, and PmPpA in the EM UDP–GlcA

We already analyzed the EM UDP–GlcNAc in our previous study [31]. Therefore, we focused in this study on the EM UDP–GlcA. Especially the enzymes AtGlcAK, AtUSP, and PmPpA were investigated for their kinetics (Table 1), optimal pH value and temperature as well as metal ion dependency (Figures S2–S4).

Table 1. Kinetic data of enzymes in the EM UDP–GlcA.

Kinetic Constants [1]	AtGlcAK	AtUSP	PmPpA
V_{max} (U/mg)	GlcA: 7.70; ATP_{app} [2]: 35.83	Glc-1-P: 129.66 UTP: 156.33	$PP_{i,\ app}$ [2]: 668000
K_m (mM)	GlcA: 0.62 ATP_{app} [2]: 8.56	Glc-1-P: 0.58 UTP: 0.44	$PP_{i,\ app}$ [2]: 3552.58
K_{iS} (mM)	GlcA: - ATP: 2.87	-	$PP_{i,\ app}$ [2]: 0.0086

[1] Kinetic constants were calculated with equation 1 and 2. [2] Apparent kinetic constants.

AtGlcAK was inhibited by ATP concentrations higher than 10 mM (Table 1 and Figure S2A,B) resulting in different kinetic constants compared to published data [23,29,30]. The optimal pH value and temperature is 7.5–9 °C, and 35–45 °C, respectively (Figure S2C,D). The temperature optimum corresponds to a prior study by Pieslinger et al., while we detected a broader pH optimum [30]. Mg^{2+} concentrations should be at least equimolar to the optimal ATP concentration of 5 mM to reach high specific enzyme activity (Figure S2E,F). For AtUSP, kinetic constants (Table 1 and (Figure S3A,B) were similar to published data [23,25,27,36]. The highest specific activity was found at a temperature of 40–45 °C (Figure S3C) and at a pH value of 8.0–8.5 (Figure S3C). Excess of Mg^{2+} over the UTP concentration results in high specific activity (Figure S3E). For PmPpA, a substrate concentration higher than 5 mM PP_i strongly inhibits the enzyme activity (Table 1 and Figure S4A). The enzyme has a temperature optimum at 40 °C (Figure S4B), a pH optimum at 8.5–9 (Figure S4D), and at least a three-fold Mg^{2+} excess to the PP_i concentration leads to high enzyme activities (Figure S4C).

2.3. *Characterization of the EM UDP–GlcA*

AtGlcAK, AtUSP, and PmPpA were combined for the one-pot synthesis of UDP–GlcA (Figure 1). UDP–GlcA synthesis reaches 80% and 84% yield after 0.5 and 1 h, respectively. We significantly improved the space-time-yield (STY) reaching 16.1 mmol L^{-1} h^{-1} for UDP–GlcA when compared to a previous study on the same enzyme cascade reporting 89% yield after 12 h and a STY of 1.5 mmol L^{-1} h^{-1} [24]. Li et al. reached 50% yield after 10 h without PmPpA [23]. PmPpA is essential for reaching high product yields and without PmPpA we observed only 32% after 24 h (Figure S5). We want to point out, that we observed a relatively high UDP concentration (1.5 mM) in the purchased UTP [37].

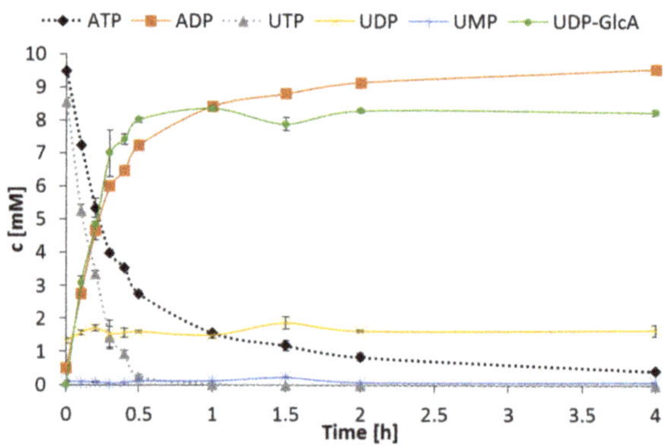

Figure 1. Synthesis of UDP–GlcA with AtGlcAK, AtUSP, and PmPpA. One-pot synthesis was performed under the following conditions: 100 mM 2-[4-(2-hydroxyethyl)piperazin-1-yl]ethanesulfonic acid (HEPES) pH 8, 25 °C, 10 mM ATP, 10 mM UTP, 10 mM GlcA, 20 mM $MgCl_2$, 59 μg/mL AtGlcAK, 54 μg/mL AtUSP and 692 μg/mL PmPpA using a volume of 300 μL. Nucleotides and nucleotide sugars were detected with multiplexed capillary electrophoresis (MP-CE).

Mn^{2+} is the preferred metal ion for PmHAS, while Mg^{2+} is the preferred metal ion for AtGlcAK and AtUSP [30,36,38]. In our previous work, we found activation of PmHAS by K^+ in the presence of Mg^{2+} or Mn^{2+} [31]. Therefore, the influence of combining the metal ions Mn^{2+}, Mg^{2+}, and K^+ on the enzyme activities and the performance of the EM UDP–GlcA was further investigated (Figure 2).

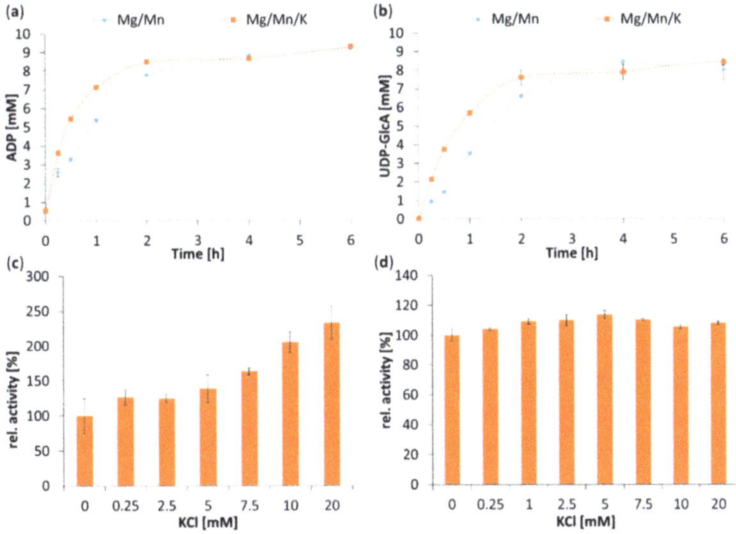

Figure 2. Influence of Mg^{2+}, Mn^{2+}, and K^+ on the EM UDP–GlcA. Course of AtGlcAK (**a**) and AtUSP (**b**) reaction within the EM UDP–GlcA with two different cofactor compositions. The reactions included 100 mM HEPES pH 8, 25 °C, 10 mM ATP, 10 mM UTP, 10 mM GlcA, 46 μg/mL AtGlcAK, 92 μg/mL AtUSP, 239 μg/mL PmPpA, and either 10 mM Mg^{2+}, and 10 mM Mn^{2+} or 10 mM Mg^{2+}, 10 mM Mn^{2+}, and 10 mM K^+. The reaction volume was 300 μL. (**c**): Activity assay for AtGlcAK: 100 mM HEPES pH 8, 25 °C, 5 mM ATP, 5 mM GlcA, 5 mM Mg^{2+}, 0–20 mM K^+, and 62 μg/mL AtGlcAK. 100% activity = 2.34 U/mg. (**d**): Activity assay for AtUSP in the presence of PmPpA: 100 mM HEPES pH 8, 25 °C, 5 mM UTP, 5 mM Glc-1-P, 5 mM Mg^{2+}, 0–20 mM K^+, 58 μg/mL AtUSP, and 18 μg/mL PmPpA using a volume of 300 μL. 100% activity = 121.46 U/mg.

All combinations of K^+ with Mg^{2+} and Mn^{2+} gave higher conversion rates of the EM UDP–GlcA within the first hour compared to the corresponding reactions without K^+ (Figure 2a,b and Figure S6A,B). After 6 h, ~90% yield for ADP and ~85% yield for UDP–GlcA were reached with and without K^+. When only Mn^{2+} or Mg^{2+} are used as a metal ion, we noticed, that Mn^{2+} has a positive effect on UDP–GlcA production (Figure S6A,B). The AtUSP/PmPpA system seems to prefer Mn^{2+}, although AtUSP activity was described to be equal for both metal ions [27]. The kinase AtGlcAK does not prefer one of these cofactors, as described in the literature [30]. Important to note is that after 24 h reactions with Mn^{2+} gave lower yields of UDP–GlcA compared to Mg^{2+} (Figure S6B). Mn^{2+} causes decomposition of UDP–GlcA into the 1,2-cyclic phosphate derivate and uridine monophosphate (UMP) (Figure S6C) [31,39].

To gain more insight, the influence of K^+ on the enzyme activity of AtGlcAK and AtUSP/PmPpA was investigated (Figure 2c,d). While the activity does not change in the AtUSP/PmPpA system in the presence of K^+, the AtGlcAK activity is increased 2.3-fold with 20 mM K^+ compared to the reaction without K^+. This is a novel feature of glucuronokinases [40,41]. PmHAS also shows higher activity in the presence of K^+ (Figure S7), confirming our previous study [31]. However, K^+ has no beneficial effect on the activities of AtUSP, BlNahK, SzGlmU, and PmPpA (Figure S7). A similar, but more intense effect is known for pyruvate kinases, where K^+ increases the affinity for ADP–Mg^{2+} [42,43]. For human pyridoxal kinase, activity is increased by K^+ [44,45]. For AtGlcAK we may assume that K^+ promotes binding of the ATP–Mg^{2+} complex or release of the phosphorylated sugar from the active site. In conclusion, all enzymes of the EM UDP–GlcA are active in the presence of Mn^{2+} and K^+. Thus, we can apply our previously published optimal conditions for good conversion rates of PmHAS and high molecular weight for HA: these are 15 mM Mg^{2+}, 10 mM K^+, and 1.5 mM Mn^{2+} [31].

2.4. One-Pot Synthesis: Combination of the EM UDP–GlcA with the EM HA

Kinetic data of PmHAS from our previous study reveal a substrate inhibition by UDP–GlcA (K_{Si} 0.57 mM) and a relatively low affinity for UDP–GlcNAc (K_m 23.4 mM) [31]. In general, a high UDP–GlcNAc/UDP–GlcA ratio with a constant supply of UDP–GlcA at low concentrations was favorable for the synthesis of 2 MDa HA. In the present study, we varied substrate concentrations in the EM UDP–GlcA at a constant UDP–GlcNAc concentration (15 mM) to investigate the effect of UDP–GlcA formation on PmHAS activity (formation of UDP) and HA synthesis (Figure 3 and Figure S8).

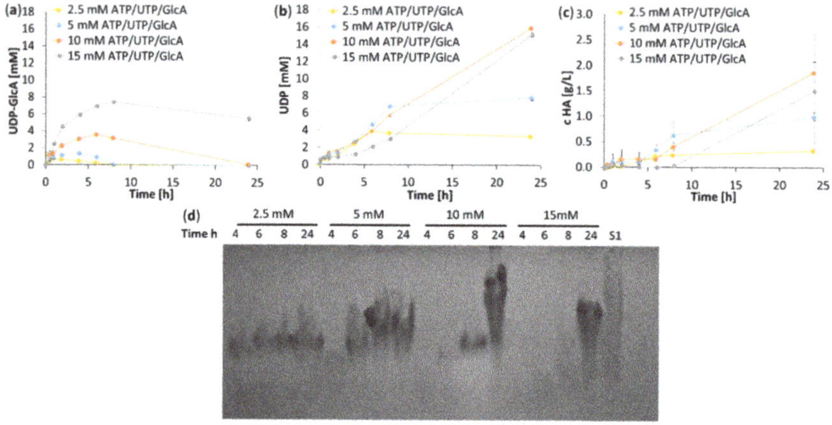

Figure 3. Combination of the EM UDP–GlcA with the EM HA. For the reaction 300 µL were placed in a 96 well plate under the following conditions: 100 mM HEPES pH 8, 25 °C, 15 mM UDP–GlcNAc, X mM ATP/UTP/GlcA, 15 mM Mg^{2+}, 10 mM K^+, 1.5 mM Mn^{2+}, 230 µg/mL AtGlcAK, 92 µg/mL AtUSP, 239 µg/mL PmPpA, and 569 µg/mL PmHAS. The course of UDP–GlcA synthesis (**a**) and UDP formation (**b**), as well as the concentration of HA determined with CTM (**c**), and the size of HA determined with agarose gel electrophoresis (**d**), are depicted. S1: high-M_w HA ≥ 2 MDa, Figure S8 depicts the concentrations of all nucleotides and nucleotide sugars during the time course of the enzymatic reactions.

At low substrate concentrations for the EM UDP–GlcA (2.5 and 5 mM) UDP–GlcA is completely converted after 8 h reaction time (Figure 3a). PmHAS reaches a high activity (0.85 µmol UDP mL^{-1} h^{-1}) as calculated from the slope of UDP formation (Figure 3b) and the highest HA size at 5 mM start concentration after 8 h (Figure 3d). At high GlcA concentrations (15 mM) HA formation is limited, the PmHAS activity decreases to 0.38 µmol UDP mL^{-1} h^{-1} and the highest UDP–GlcA concentration of 8 mM after 8 h is reached. At 10 mM substrate concentration, the activity of PmHAS still reaches 0.72 µmol UDP mL^{-1} h^{-1}, however, HA size is low after 8 h. In conclusion, these data reflect the activity of PmHAS being high at UDP–GlcA concentrations ≤ 5 mM with immediate HA formation within 8h. At higher UDP–GlcA concentrations, PmHAS activity reaches a limit and shows delayed HA formation. This limit is reached with 10 mM substrate concentrations in the EM UDP–GlcA. Using 10 mM of GlcA, UTP, and ATP, UDP–GlcA is completely converted after 24 h resulting in 1.8 g/L HA and the highest HA size. As a compromise between PmHAS activity, high molecular weight, and concentration after 24 h, 10 mM starting concentrations of substrates in the EM UDP–GlcA was used for the next experiment.

2.5. One-Pot Synthesis: Combination of EM UDP–GlcA and EM UDP–GlcNAc with EM HA

We combined both nucleotide sugar modules with the HA module in a volume of 5 mL and started the one-pot synthesis with 10 mM GlcA and GlcNAc, respectively (Figure 4). Within 3 h reaction time, UTP is nearly completely converted resulting in ~10 mM UDP–GlcNAc and ~6 mM UDP–GlcA (Figure 4a) starting the production of HA (Figure 4b,c). These concentrations are favorable for an optimal reaction rate of PmHAS (1.26 µmole mL^{-1} h^{-1} UDP) as calculated from the slope of UDP formation (Figure 4a). Interestingly, the size of the HA chain (median values for HA chains' M_w) gradually increases to M_w 0.86 MDa within 8 h reaction time (Figure 4d). It is possible to stop the reaction at certain time points to get the preferred HA size. However, the concentration levels of HA are relatively low (Figure 4b). That is also the reason why we could not measure the size of HA before 4 h, although HA is already produced (Figure 4c). Within 8 h, 0.28 g/L HA with an average M_w of 0.86 MDa and dispersity of 1.11 is reached. After 24 h, the final HA concentration was ~1.4 g/L with an average M_w of 0.7 MDa and dispersity of 1.16. Interestingly, the maximum HA chain length ranges between 1.54 MDa after 8 h and 2.2 MDa after 24 h. In addition, MP-CE analysis reveals that ~1 mM UMP is formed and ~1 mM UDP–GlcNAc is not utilized after 24 h (Figure 4a). Mn^{2+} degrades UDP–GlcA into a 1,2-cyclic phosphate derivate and UMP. UDP–GlcNAc is stable under these conditions [39]. In summary, ~1 mM UDP–GlcA is not available for the HA synthesis and, because of stoichiometric reasons, 1 mM UDP–GlcNAc are left. Thus 2 mM from 20 mM UTP are not used for the HA production resulting in 90% yield. One way to increase the yields is by varying the Mn^{2+} concentration, although it is already adjusted to the PmHAS activity [31]. The other way is to decrease the pH value because the decomposition is more intense in a basic environment [39]. Compared with the work of Li et al., we reached a higher yield (90% vs. 70%) in shorter reaction time (24 vs. 40 h) [23]. A maximum HA chain length of 0.55 MDa was obtained [23]. These differences are probably due to a different approach to synthesize HA, e.g., equal concentrations of UDP–GlcA and UDP–GlcNAc were applied resulting in a lower activity of PmHAS [31].

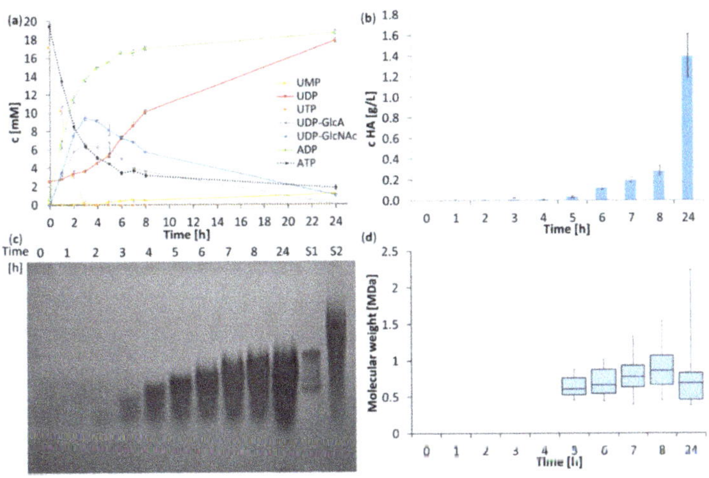

Figure 4. Combination of EM UDP–GlcA and EM UDP–GlcNAc with the EM HA. Conditions: 100 mM HEPES pH 8, 25 °C, 10 mM GlcA, 10 mM GlcNAc, 20 mM ATP, 20 mM UTP, 15 mM Mg^{2+}, 10 mM K^{+}, 1.5 mM Mn^{2+}, 108 µg/mL AtGlcAK, 11 µg/mL AtUSP, 108 µg/mL BlNahK, 109 µg/mL SzGlmU, 46 µg/mL PmPpA, and 1191 µg/mL PmHAS. A volume of 5 mL was used. (**a**): Time course of nucleotide and nucleotide sugar concentrations. (**b**). HA concentration analyzed with CTM. (**c**): HA size analysis by agarose gel electrophoresis with standards S1: Select-HA™ HiLadder (Hyalose LLC) with a composition of defined HA preparations (495, 572, 966, 1090, and 1510 kDa), high-M_w HA ≥ 2 MDa, S2: high-M_w HA ≥ 2 MDa. (**d**): Quantitative measurement of the HA-size with SEC-RALS/LALS. Distribution of HA chains' Mw is presented as a box plot diagram. The box represents 50% of the HA chains, while each whisker represent 25% of HA chains. The line within the box indicates the median value of HA chains' M_w.

2.6. Influence of pH on the One-Pot Synthesis of HA

To investigate further the influence of the UDP–sugar ratio on the PmHAS activity in this one-pot synthesis, we varied two system parameters: pH and Mg^{2+}. Control of UDP–sugar formation should lead to an optimal HA production with control over HA size. Variation of the pH value shall be a key factor to adjust UDP–sugar ratios since the EM UDP–GlcA and EM UDP–GlcNAc perform differently at specific pH values (Figure 5a and Figure S9). Especially the activity of AtGlcAK is reduced at a pH value higher than 8 (Figure S2D), while the EM UDP–GlcNAc has a broad pH spectrum [31]. After 2 h 10 mM UDP–GlcNAc are produced at pH 8.5 or 9 (Figure 5a), while the UDP–GlcA concentration is lower (5–7 mM). High UDP–GlcNAc/UDP–GlcA ratios lead to higher PmHAS activity as demonstrated for the high UDP yield in Tris-HCl pH 9 (68.8%) after 8 h (Figure S10) reaching final yields of 82% and 86% at pH 8.5 and 9.0, respectively (Figure 5b). These data reflect the relative broad pH optimum of PmHAS as demonstrated in our previous study [31]. As we mentioned earlier, one way to increase the yield could be the decrease of the pH value to reduce the decomposition of UDP–GlcA by Mn^{2+}. However, a decrease of the pH value has a greater effect on the functionality of the enzymes (Figure 5b). Final HA concentrations differ for the different buffers at pH values between pH 8 and pH 8.5–9 (Figure 5c). High HA concentrations were obtained at pH 8 with Tris-HCl (1.95 g/L), at pH 8.5 with 2-[4-(2-hydroxyethyl)piperazin-1-yl]ethanesulfonic acid HEPES-NaOH (2.68 g/L), and at pH 9 with 3-(Cyclohexylamino)-1-propanesulfonic acid (CAPSO)-NaOH (2.54 g/L). The size of the HA is also dependent on the pH value (Figure 5d,e). The qualitative analysis showed that the largest HA chains are produced at a pH value of 8.5 and 9 (Figure 5d). This result is confirmed by size exclusion chromatography, showing the highest M_w with Tris-HCl at pH 8 (1.31 MDa), pH 8.5 (1.50 MDa), and pH 9 (1.38 MDa) (median values for HA chains' M_w in Figure 5e and Table S1). The dispersity of these samples in Tris-HCl is between 1.1 and 1.3 (Table S1). We conclude that the best buffer and pH value is HEPES or Tris-HCl at pH 8.0–8.5 for one-pot synthesis of HA in an M_w range of 1.3–1.5 MDa. Setting this system parameter, the beneficial UDP–GlcA/UDP–GlcNAc ratio in the starting phase of the one-pot synthesis adjusts PmHAS activity for HA polymerization.

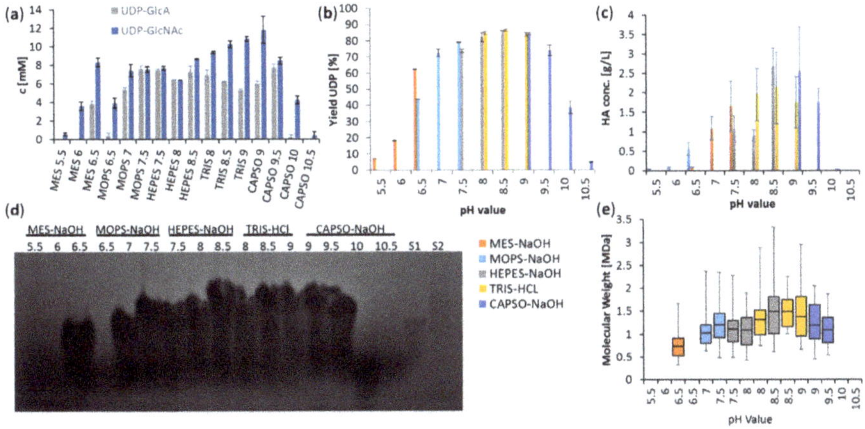

Figure 5. Controlling HA one-pot synthesis with pH value. 100 mM buffer pH X, 25 °C, 10 mM GlcA, 10 mM GlcNAc, 20 mM ATP, 20 mM UTP, 15 mM $MgCl_2$, 10 mM KCl, 1.5 mM $MnCl_2$, 124 µg/mL AtGlcAK, 117 µg/mL AtUSP, 1202 µg/mL BlNahK, 255 µg/mL SzGlmU, 369 µg/mL PmPpA, 1289 and µg/mL PmHAS. A volume of 5 mL was used. Time course of the reaction of all pH conditions analyzed with MP-CE. All graphs can be found in the supplemental data (Figure S9). (**a**): Concentrations of nucleotide sugars UDP–GlcA and UDP GlcNAc after 2 h. (**b**): Yield of the reaction after 24 h are calculated by the generated UDP concentration and the UTP starting concentration. (**c**): HA concentration after 24 h measured with CTM. (**d**): Qualitative analysis of the HA-size with agarose gel electrophoresis after 24 h. S1: Select-HA™ HiLadder (Hyalose LLC) with a composition of defined HA preparations (495, 572, 966, 1090, and 1510 kDa), S2: high-M_w HA ≥ 2 MDa. (**e**): Quantitative measurement of the HA-size with Size-Exclusion Chromatography with Right-Angle/Low-Angle Light Scattering (SEC-RALS/LALS). Distribution of HA chains' *Mw* is presented as a box plot diagram. The box represents 50% of the HA chains, while each whisker represent 25% of HA chains. The line within the box indicates the median value of HA chains' M_w.

2.7. Influence of Magnesium on the One-Pot Synthesis of HA

In our previous study metal cofactor concentrations were set at 15 mM Mg^{2+}, 1.5 mM Mn^{2+} and 10 mM K^+ for optimal HA synthesis [31]. In the present study using the EM UDP–GlcA, we varied the Mg^{2+} concentration to find the optimal system parameter settings for this one-pot synthesis. (Figure 6 and Figure S11). Most interestingly, a variation of Mg^{2+} concentration has a significant influence on the UDP–GlcNAc/UDP–GlcA ratio (Figure 6a). With 20 mM and 25 mM, a high ratio can be adjusted after 2 h reaction time. At a low Mg^{2+} level (up to 10 mM Mg^{2+}), UDP–GlcA is in excess of UDP–GlcNAc. At 15 mM Mg^{2+}, both nucleotide sugars are synthesized and converted in equal amounts. These ratios are observed during the entire reaction time (Figure S11) and significantly affect HA synthesis. The ratio of both nucleotide sugars controls the activity of PmHAS as seen in the UDP yield (Figure 6b), HA titer (Figure 6c) as well as HA M_w (Figure 6d,e, Table S2). While for PmHAS the UDP–GlcNAc concentration should be as high as possible, the optimal UDP–GlcA concentration lays between 3 and 7 mM [31]. The best ratio is reached with 25 mM Mg^{2+} and results in the highest UDP yield (80.2%) (Figure 6b), a high HA concentration (1.49 g/L) (Figure 6c) and the largest HA size (median value for HA chains' M_w = 1.55 MDa) (Figure 6e) with an excellent dispersity of 1.049 (Table S2). In contrast, PmHAS activity is lower with an excess of UDP–GlcA over UDP–GlcNAc and results in the formation of lower molecular size HA. Most important, each Mg^{2+} concentration gives a distinct HA M_w with excellent dispersity and good yield (Figure 6b–e, Table S2). In conclusion, with defined Mg^{2+} concentration in one-pot HA synthesis, we adjust PmHAS activity by the obtained ratios of UDP–GlcNAc and UDP–GlcA and in this way HA size and titer.

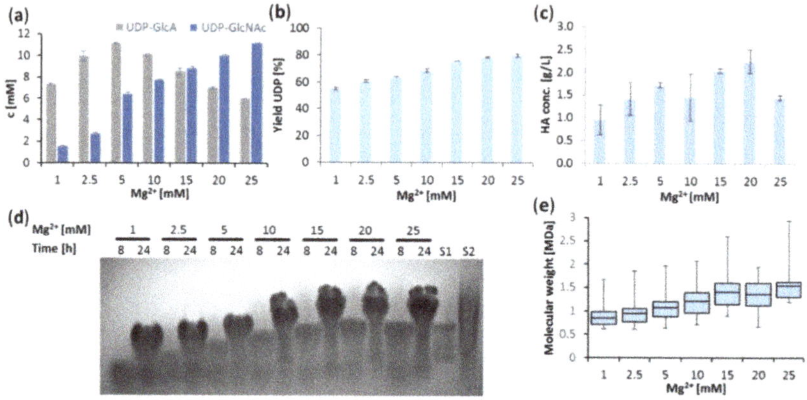

Figure 6. Controlling HA one-pot synthesis with Mg^{2+} concentration. 100 mM HEPES–NaOH pH 8, 25 °C, 10 mM GlcA, 10 mM GlcNAc, 20 mM ATP, 20 mM UTP, X mM $MgCl_2$, 10 mM KCl, 1.5 mM $MnCl_2$, 1240 µg/mL AtGlcAK, 117 µg/mL AtUSP, 1202 µg/mL BlNahK, 255 µg/mL SzGlmU, 369 µg/mL PmPpA, and 1289 µg/mL PmHAS. A volume of 5 mL was used. (**a**): Ratio of the intermediate nucleotide sugars UDP–GlcA and UDP–GlcNAc after 2 h. Nucleotides and nucleotide sugars measured with MP-CE. Graphs can be found in the supplemental data (Figure S11). (**b**): Yield of the reaction calculated with the generated UDP concentration and used UTP starting concentration. (**c**): HA concentration after 24 h measured with CTM. (**d**): Qualitative analysis of the HA-size with agarose gel electrophoresis after 8 and 24 h. S1: Select-HA™ HiLadder (Hyalose LLC) with a composition of defined HA preparations (495, 572, 966, 1090, and 1510 kDa), S2: high-M_w HA ≥ 2 MDa. (**e**): Quantitative measurement of the HA-size with SEC-RALS/LALS. Distribution of HA chains' M_w is presented as a box plot diagram. The box represents 50% of the HA chains, while each whisker represent 25% of HA chains. The line within the box indicates the median value of HA chains' M_w.

Comparing the pH value and the Mg^{2+} concentration as control parameters in the one-pot synthesis of HA, we have different kinds of scenarios. Setting the pH value at pH 8.5–9.0, we adjust a good ratio of the UDP–sugars, but PmHAS does not perform at its pH optimum [31,38]. At 25 mM Mg^{2+} concentration, a beneficial UDP–sugar ratio is reached, and also the activity of PmHAS increases with higher Mg^{2+} concentrations in the presence of K^+ and Mn^{2+}. In both cases, it seems that excess of UDP–GlcNAc over UDP–GlcA gives higher HA yields. Moreover, control over HA polymer size is gained by adjusting the UDP–sugar ratio, preferentially by pH and Mg^{2+} concentration. A similar effect was reported for the microbial HA production with *S. zooepidemicus*, where the addition of GlcNAc or overexpression of genes involved in UDP–GlcNAc synthesis resulted in higher molecular weight of HA, whereas overexpression of UDP–GlcA genes had the opposite effect [46–48]. The HAS enzyme in *S. zooepidemicus* is classified as a class 1 type, whereas PmHAS is a class 2 type. [4,18,49]. Both types differ in form, membrane linkage and catalytic reaction [50]. Nevertheless, both enzymes may share biochemical features for HA production and having adapted similarly to their environment [51]. Previous studies showed that PmHAS mode of action could be divided into de novo and polymerization synthesis, where the polymerization step is much faster [13,15]. Studies showed that the GlcA-transferase domain needs at least HA_4 and the GlcNAc-transferase domain needs at least HA_3 oligosaccharides for polymerization speed [52]. Interestingly the GlcA-transferase activity is enhanced with longer HA oligosaccharides [22]. HA chain length and polydispersity are greatly dependent on the amount of de novo synthesized chains in the early stages of the reaction. The more HA chains are created, the more the UDP–sugars are distributed among the HA chains. This results in shorter polymers [13,15]. It is possible that a certain UDP–sugar ratio could influence the preference of PmHAS for polymerization or de novo synthesis at the beginning of the reaction, which then would result in different HA sizes after 24 h.

2.8. Comparison with Industrial Production Processes

Depending on the demand, the size of the HA chain can be adjusted with the in vitro one-pot synthesis through the UDP–GlcA/UDP–GlcNAc ratio. Another way to regulate the size is, to stop the reaction earlier. The dispersity for all shown reactions is relatively low. With HA extraction from animal tissue, size and dispersity are very hard to control. Both characteristics are hampered trough intrinsically low yields and high dispersity within the tissue and endogenous hyaluronidase activity [53]. Also with bacterial fermentation, low dispersity is still a challenge and is dependent on culture conditions [12,54]. On top of that, extraction from animal tissue and production with pathogenic *Streptococcus* strains bears the risk of biological contamination, which increases the effort and cost of product purification. [3]. The named issues would cease with this in vitro approach.

With the conditions of 15 mM Mg^{2+} and HEPES–NaOH pH 8.5, we could reach an M_w of 1.49 MDa with a final HA concentration of 2.7 g/L and with 20 mM Mg^{2+} and HEPES–NaOH pH 8 we could reach 1.36 MDa with a final HA concentration of 2.2 g/L. These characteristics can compete with results from bacterial fermentation [55]. However, in our approach, the reaction volume of 5 mL is only lab scale. A scale-up will be necessary for HA production and therefore the preparation of the enzymes will become important [53]. Considering that purification of enzymes is labor and cost-intensive, good long-term storage stability is favorable. Therefore, in the present work the activity of every enzyme was evaluated over four weeks (Figure S7). All enzymes were stored in 100 mM HEPES pH 8 at 4 °C. BlNahK, SzGlmU, and PmHAS are very stable. The AtUSP activity was reduced by ~55% after three weeks. AtGlcAK and PmPpA are most crucial by losing ~70% and ~64% of their activity after one week, respectively. Due to the very high initial activity of PmPpA, this residual activity is still sufficient for HA synthesis. However, the activity loss of AtGlcAK is an issue that will need to be addressed. To enhance the storage stability, immobilization or addition of additives can be used [56]. For example, sucrose enhances the stability of some enzymes significantly [31]. Protein engineering could also help to improve stability or other characteristics as it was recently shown that a PmHAS variant with better activity and chain length specificity was created [57].

3. Materials and Methods

Nucleotides and nucleotide sugars are from Carbosynth Limited UK (Compton, UK). All other chemicals are from Carl Roth GmbH (Karlsruhe, Germany), if not further mentioned.

3.1. Enzyme Production

3.1.1. Cloning of Recombinant Genes

The synthetic genes of GlcAK from *Arabidopsis thaliana* (Gene ID: 819902), USP from *Arabidopsis thaliana* (Gene ID: 835333) and PpA from *Pasteurella multocida* (Gene ID: 29387852) were modified with restriction recognition sequences (Table 2).

Table 2. Modification of recombinant genes.

Gene	N-Terminal	C-Terminal
glcak	*Nde*I GGAATTCCATATGGAATTCC	*Xho*I CCGCTCGAGCGG
usp	*Xho*I CCGCTCGAGCGG	*BamH*I CGGGATCCCG
ppa	*Nde*I GGAATTCCATATGGAATTCC	*Xho*I CCGCTCGAGCGG

The start codons were removed from the sequence of *glcak* and *ppa* because of the attachment of the *Nde*I sequence, in which is already a start codon. The stop codons were also removed to clone this sequence into the pET-22b(+) vector with C-terminal His_6-tag. The start codon of *usp* was removed to clone this gene into pET-16b(+) with N-terminal His_{10}-tag. All genes were optimized for *Escherichia coli* codon usage and were ordered from GenArt™. The genes were cloned into pET-22b(+) and pET-16b(+), respectively. The cloning was confirmed by sequencing and restriction analysis. The genes for BlNahK from *Bifidobacterium longum* (GenBank accession number: AB303839.1), SzGlmU from *Streptococcus zooepidemicus* (GenBank accession number: AF347022) and $HAS^{1\text{-}703}$ from *Pasteurella multocida* (GenBank accession number: AF036004.2) were already cloned into pET-22b(+) vectors in our previous work [31].

3.1.2. Transformation and Cultivation

All constructs pET-22b-*atglcak*, pET-16b-*atusp*, pET-22b-*pmppa*, pET-22b-*blnahk*, pET22b-*szglmu* and pET-22b-$pmhas^{1\text{-}703}$ were transformed into *E. coli* BL21 (DE3) via heat shock. The transformed clones were selected on agar plates with lysogeny broth (LB) medium containing ampicillin (100 µg/mL). Clones were further cultivated in baffled shaking flasks. The pre-culture was grown in 20 mL LB medium with ampicillin (37 °C, 120 rpm). The main culture (1 L) with terrific broth was inoculated with 1% (*v*/*v*) pre-culture (37 °C, 80 rpm). When the OD_{600} of the main culture reached 0.6–0.8, the expression of the proteins was induced by adding 0.1 mM isopropyl β-D-1-thiogalactopyranoside (IPTG, AppliChem GmbH, Darmstadt, Germany) and the temperature was reduced to 25 °C. After 20 h the cells were centrifuged (7000 rpm, 4 °C), the supernatant was discarded and the cells were stored at −20 °C.

3.1.3. Enzyme Purification

Frozen cells (4 g) were suspended with 10 mL binding buffer (20 mM sodium phosphate, 500 mM sodium chloride, 30 mM imidazole, pH 7.4), disrupted with ultrasound and subsequently centrifuged (30 min, 15,000 rpm, 4 °C). The recombinant enzymes containing a His-Tag were purified by immobilized metal ion affinity chromatography (IMAC) using 5 mL HisTrap HP columns from GE Healthcare and the Äktapurifier 100 controlled with the program Unicorn. Target enzymes were eluted using elution buffer (20 mM sodium phosphate, 500 mM sodium chloride, 500 mM imidazole, pH 7.4). The elution buffer was changed by dialysis with 100 mM HEPES pH 8. Protein concentration was measured by the Bradford protein assay. The purification was confirmed by sodium dodecyl sulfate polyacrylamide gel electrophoresis (SDS-PAGE) and Western blot using a His_6-tag monoclonal antibody, which is conjugated with horseradish peroxidase (Roche Diagnostics GmbH, Mannheim, Germany). The same method for PmHAS and the enzymes of the EM UDP–GlcNAc was used as described in our previous study [31].

3.2. Enzyme Assays

3.2.1. Activity Assays

All enzyme activities were tested in a reaction volume of 300 µL in a 96 well plate at 25 °C and pH 8 (100 mM HEPES–NaOH). The composition of each enzyme reaction is listed as follows: AtGlcAK: 5 mM ATP, 5 mM GlcA and 10 mM $MgCl_2$. AtUSP: 5 mM UTP, 5 mM Glc-1-P, 10 mM $MgCl_2$. Glc-1-P was used instead of GlcA-1-P because of cost reasons. Glc-1-P is hardly available. BlNahK: 5 mM ATP, 5 mM GlcNAc and 10 mM $MgCl_2$. SzGlmU: 5 mM UTP, 5 mM GlcNAc-1-P, 10 mM $MgCl_2$. PmHAS: 10 mM UDP–GlcA, 10 mM UDP–GlcNAc, 10 mM $MnCl_2$. PmPpA: 5 mM PP_i, 10 mM $MgCl_2$. For K^+ experiments, 10 mM KCl was added. Several samples were taken within 10 min. For PmHAS, several samples were taken within 6 h. Reactions were stopped with a solution consisting of SDS, *para*-aminobenzoic acid (PABA, Sigma Aldrich, Munich, Germany) and *para*-aminophthalic acid (PAPA, Sigma Aldrich) in a ratio of 1:1 resulting in a final concentration of 7 mM SDS, 1 mM PABA and 1 mM PAPA. The reactions for AtGlcAK, AtUSP, BlNahK, SzGlmU, and PmHAS were analyzed with multiplexed capillary electrophoresis and for PmPpA with a phosphate assay kit (MAK308, Sigma Aldrich, Munich, Germany). For the kinases ADP, for AtUSP UDP–GlcA, for SzGlmU UDP–GlcNAc, for PmHAS UDP, and for PmPpA P_i as products were monitored. The activity (U = µmol/min) was determined via the slope during the linear phase and the volumetric activity (U/mL) was calculated. With the enzyme concentration from the Bradford assay, the specific activity (U/mg) was calculated.

3.2.2. Kinetic Assays

Kinetics for AtGlcAK, AtUSP, and PmPpA were assayed in a volume of 300 µL in a 96 well plate. The following conditions were used for AtGlcAK: 25 °C, 100 mM HEPES pH 7.5 and 5 mM $MgCl_2$. ATP or GlcA, respectively, was set to a constant concentration of 5 mM, while the other one was set in a range from 0.5–30 mM. Kinetics for AtUSP were tested with 5 mM UTP or 5 mM Glc-1-P, while the other substrate was varied between 0.25–25 mM. Conditions were 25 °C, 100 mM HEPES pH 8 and 15 mM $MgCl_2$. PmPpA: 25 °C, 100 mM HEPES pH 8. $MgCl_2$ concentrations were adjusted to PP_i concentrations (1:1). PP_i ranged from 1–40 mM. Samples were taken at serval times within 10 min. Reactions were stopped as described for the activity tests. The kinetic parameters were determined with the Michaelis-Menten equation (Equation (1)) or with substrate inhibition (uncompetitive) equation (Equation (2)) using the program SigmaPlot 10.0 (SPSS Software GmbH, Erkrath, Germany).

$$v = V_{\max} \cdot \frac{[S]}{K_M + [S]} \tag{1}$$

$$v = V_{\max} \cdot \frac{1}{1 + \frac{K_M}{[S]} + \frac{[S]}{K_{iS}}} \tag{2}$$

[S]: Concentration of the substrate (mM). K_M: Michaelis constant (mM). v: Rate of formation of product (µmol min^{-1} mg^{-1}). V_{max}: Maximal rate of the system (µmol min^{-1} mg^{-1}). K_{iS}: Dissociation constant of the enzyme-substrate-substrate complex.

3.2.3. One-Pot Syntheses with the EM UDP–GlcA and Whole Cascade

All reactions were tested with 300 µL in a 96 well plate or with 5 mL in a 5.5 mL reaction vessel. All reactions took place at 25 °C. Conditions vary depending on the one-pot synthesis and are directly described in the caption of the figures. Buffers, substrates, cofactors were prepared and a sample of this master mix served as the initial state. Enzymes were added and similar to the activity assays samples were taken at certain time points. Reactions were stopped 1:1 with a solution of SDS, PABA, and PAPA (see 3.2.1 activity assays). Samples were analyzed with MP-CE and if HA was produced, agarose gel electrophoresis, Cetyltrimethylammonium bromide (CTAB) turbidimetric, and SEC-RALS/LALS assays were performed.

3.3. *Analysis of Enzymatic Syntheses*

3.3.1. Multiplexed Capillary Electrophoresis (MP-CE) Analysis

All enzymatic reactions (except for PmPpA) were analyzed by a multiplexed cePRO 9600™ system (Advanced Analytical Technologies, Ames, IA, USA) [35]. The separation of nucleotides and nucleotide sugars for 96 samples was performed in a 96-capillary array, where each capillary had an effective length of 55 cm, a total length of 80 cm and an inner diameter of 50 µm. 10 kV voltage was applied with an electrophoresis buffer (70 mM ammonium acetate, 1 mM EDTA, pH 9.2). Samples were injected by vacuum (−0.7 psi, 10 s). PABA and PAPA were used as an internal standards with a concentration of 1 mM each. The analytes were detected by UV measurement (254 nm) and the resulting data were evaluated with the pK_a-Analyzer software (Advanced Analytical Technologies, Ames, IA, USA). The amount of the analytes were calculated by means of the peak areas.

3.3.2. Phosphate Assay Kit

The enzymatic syntheses with PmPpA were analyzed by a phosphate assay kit (Sigma-Aldrich, MAK308) in accordance with the manufacturer's instructions. The standard curve ranged from

4–40 μM P_i made from a 1 mM P_i stock solution. The test sample was diluted to fit in the standard range; 50 μL diluted sample was mixed with 100 μL of the malachite green reagent in a 96 well plate and incubated for 30 min at room temperature. The absorbance was measured at 620 nm.

3.4. Analysis of Hyaluronic Acid

3.4.1. Agarose Gel Electrophoresis

The molecular weight of hyaluronic acid was analyzed by agarose gel electrophoresis as described in the literature [58]. Samples and standards were mixed with 4 μL loading buffer (ThermoFisher scientific) and applied on a 0.5% agarose gel. The separation was done with 105 V for 55 min in TAE buffer (40 mM TRIS acetate, 1 mM EDTA, pH 8). In the dark, the gel was washed 30 min with 30% ethanol, dyed for 24 h with Stains-All solution (Sigma-Aldrich, 30% ethanol, 6.25 μg/mL Stains-All), discolored with water and documented at the GelDoc (BioRad, Düsseldorf, Germany). Two markers were used: (a) Select-HA™ HiLadder (Hyalose LLC) with a composition of defined HA preparations (495, 572, 966, 1090 and 1510 kDa), (b) high-M_w HA ≥ 2 MDa (Contipro Biotech, Dolní Dobrouč, Czech Republic).

3.4.2. CTAB Turbidimetric Method (CTM)

CTM is a method to measure the concentration of HA, which is based on the formation of turbidity between HA and cetyltrimethylammonium bromide (CTAB) [59–61]. 25 μL of acetate buffer (200 mM sodium acetate, 150 mM sodium chloride, pH 6) were mixed with 25 μL sample (1:20 dilution). CTAB solution of 100 μL (25 g/L CTAB in 2% *w/v* sodium hydroxide) was added, incubated for 5 min at room temperature and the readings were taken at 400 nm. From all absorption values, the associated $t = 0$ min value was subtracted and the HA concentration was calculated with linear regression. The standard curve was provided by dilutions of high-M_w HA (Contipro Biotech, $M_w \geq 2$ MDa) ranging from 6.25–200 μg/mL.

3.4.3. Size-Exclusion Chromatography with Right-Angle/Low-Angle Light Scattering (SEC-RALS/LALS)

SEC-RALS/LALS (Viscotek TDA 305, Malvern Instruments, Kassel, Germany) was used to determine the molecular weight distribution (M_w and M_n) and relative molecular mass dispersity within the HA samples, by using two linearly connected columns (A6000M, A7000, Malvern Instruments) for separation. Samples were diluted with 10 mM PBS (pH 7.4) up to a volume of 180 μL. 80 μL of this dilution was injected and were eluted with PBS at a flow rate of 0.35 mL/min and 35 °C. HA concentrations were determined by RI-detector. PEO standard (0.5 mg/mL in PBS, Malvern Instruments) was used for the calibration. The OmniSEC software employs triple detection (RI, RALS/LALS) and allows for a one-point calibration. We analyzed each sample twice and results are the mean value of two runs. The dispersity was calculated with the quotient of M_w and M_n (Equation (5)).

$$\mathrm{Mn} = \frac{\sum ni \cdot Mi}{\sum ni} \tag{3}$$

$$Mw = \frac{\sum mi \cdot Mi}{\sum mi} \tag{4}$$

$$DI = \frac{\mathrm{Mw}}{\mathrm{Mn}} \tag{5}$$

M_i: Molar mass of chains with i units. n_i: Number of chains with i units. m_i: Total mass of chains with i units. DI: Dispersity index

4. Conclusions

In this study, we aimed to understand and control the key factors for complex one-pot synthesis of HA. The one-pot synthesis contains six different enzymes and the HA chain is built up from the cheap monosaccharides GlcA and GlcNAc. As an advanced development, we present an optimized one-pot synthesis including the insights we got in our previous study [31]. For the one-pot synthesis, we used a specific metal ion composition of Mg^{2+}, Mn^{2+}, and K^+, adjusted for the combination of PmHAS and the EM UDP–GlcNAc. Surprisingly, we found out that K^+ enhances also the activity of AtGlcAK, a novel feature for this plant glucuronokinase. The pH value and Mg^{2+} concentration are key factors to control HA synthesis. Depending on these key factors, we could indeed tailor HA size, yield, and dispersity. We obtained high molecular weight HA (1.54 MDa) with low dispersity (1.05) at high (25 mM) Mg^{2+} concentrations in HEPES–NaOH buffer (pH 8). The yield was 80.2% with a final HA concentration of 1.4 g/L. High HA concentration (2.7 g L^{-1}) is reached with 15 mM Mg^{2+} in HEPES–NaOH (pH 8.5) with a yield of 86.3%, an average size of 1.49 MDa and a dispersity of 1.2. Future development should include scale-up, where immobilization of the enzymes may become important. In addition, the development of a regeneration system for the nucleotides ATP and UTP is demanding to reduce costs [62,63]. In conclusion, we demonstrated that manufacturing of high M_w and low dispersity HA product can be tailored with an in vitro one-pot synthesis.

Supplementary Materials: Supplementary materials can be found at http://www.mdpi.com/1422-0067/20/22/5664/s1.

Author Contributions: Conceptualization, L.E. and J.G.; methodology, J.G., H.Z., and A.E.; investigation, J.G. and H.Z.; data curation, J.G. and H.Z.; writing—original draft preparation, J.G. and L.E.; writing—review and editing, J.G., H.Z., J.K., A.E., and L.E.; supervision, L.E.; funding acquisition, L.E., J.K.

Funding: Financial support by the German Federal Ministry of Education and Research (BMBF) in the frame of the KMU-Innovative-17 project: "Multi-Enzym-Membranreaktor für die Synthese von hochmolekularen Hyaluronsäure-Polymeren für Kosmetik und Medizin" (031B0104B) is gratefully acknowledged.

Acknowledgments: The authors thank Truc Pham and Dennis Hirtz for their valuable contributions.

Conflicts of Interest: The authors declare no conflict of interest.

Abbreviations

ADP	Adenosine diphosphate
AMP	Adenosine monophosphate
AtGlcAK	Glucuronic acid kinase from *Arabidopsis thaliana*
ATP	Adenosine triphosphate
AtUSP	UDP–sugar pyrophosphorylase from *Arabidopsis thaliana*
BlNahK	GlcNAc-1-phosphate kinase from *Bifidobacterium longum*
CAPSO	3-(Cyclohexylamino)-1-propanesulfonic acid
CjGlmU	UDP–GlcNAc pyrophosphorylase from *Campylobacter jejuni*
CTAB	Cetyltrimethylammonium bromide
CTM	CTAB turbidimetric method
EM	Enzyme module
GlcA	Glucuronic acid
GlcNAc	*N*-acetylglucosamine
HA	Hyaluronic acid
HCl	Hydrogen chloride
HEPES	2-[4-(2-Hydroxyethyl)piperazin-1-yl]ethanesulfonic acid
IMAC	Immobilized metal affinity chromatography
IPTG	Isopropyl β-D-1-thiogalactopyranoside
K^+	Potassium cation
MES	2-Morpholin-4-ylethanesulfonic acid

Mg^{2+}	Magnesium cation
Mn^{2+}	Manganese cation
MOPS	3-Morpholinopropane-1-sulfonic acid
MP-CE	Multiplexed capillary electrophoresis
M_W	Molecular weight
NaOH	Sodium hydroxide
PABA	*para*-Aminobenzoic acid
PAPA	*para*-Aminophthalic acid
PBS	Phosphate-buffered saline
P_i	Inorganic phosphate
PmHAS	Hyaluronan synthase from *Pasteurella multocida*
PmPpA	Pyrophosphatase from *Pasteurella multocida*
PP_i	Inorganic Pyrophosphate
SDS-PAGE	Sodium dodecyl sulfate polyacrylamide gel electrophoresis
SEC-RALS/LALS	Size-Exclusion Chromatography with Right-Angle/Low-Angle Light Scattering
SzGlmU	UDP–GlcNAc pyrophosphorylase from *Streptococcus zooepidemicus*
TRIS	2-Amino-2-(hydroxymethyl)propane-1,3-diol
UDP	Uridine diphosphate
UMP	Uridine monophosphate
UTP	Uridine triphosphate
UV	Ultraviolet

References

1. Laurent, T.C.; Laurent, U.B.G.; Fraser, J.R.E. Functions of hyaluronan. *Ann. Rheum. Dis.* **1995**, *54*, 429–432. [CrossRef] [PubMed]
2. Fraser, J.R.E.; Laurent, T.C.; Laurent, U.B.G. Hyaluronan: Its nature, distribution, functions and turnover. *J. Intern. Med.* **1997**, *242*, 27–33. [CrossRef] [PubMed]
3. Fallacara, A.; Baldini, E.; Manfredini, S.; Vertuani, S. Hyaluronic acid in the third millennium. *Polymers* **2018**, *10*, 701. [CrossRef] [PubMed]
4. DeAngelis, P.L. Hyaluronan synthases: Fascinating glycosyltransferases from vertebrates, bacterial pathogens, and algal viruses. *Cell. Mol. Life Sci.* **1999**, *56*, 670–682. [CrossRef] [PubMed]
5. Parashar, P.; Rathor, M.; Dwivedi, M.; Saraf, S.A. Hyaluronic acid decorated naringenin nanoparticles: Appraisal of chemopreventive and curative potential for lung cancer. *Pharmaceutics* **2018**, *10*, 33. [CrossRef] [PubMed]
6. Cooper, C.; Rannou, F.; Richette, P.; Bruyère, O.; Al-Daghri, N.; Altman, R.D.; Brandi, M.L.; Collaud Basset, S.; Herrero-Beaumont, G.; Migliore, A.; et al. Use of intraarticular hyaluronic acid in the management of knee osteoarthritis in clinical practice. *Arthritis Care Res.* **2017**, *69*, 1287–1296. [CrossRef] [PubMed]
7. Maytin, E.V. Hyaluronan: more than just a wrinkle filler. *Glycobiology* **2016**, *26*, 553–559. [CrossRef] [PubMed]
8. Gold, M.H. Use of hyaluronic acid fillers for the treatment of the aging face. *Clin. Interv. Aging* **2007**, *2*, 369–376. [CrossRef] [PubMed]
9. Stern, R.; Asari, A.A.; Sugahara, K.N. Hyaluronan fragments: An information-rich system. *Eur. J. Cell Biol.* **2006**, *85*, 699–715. [CrossRef] [PubMed]
10. Hemshekhar, M.; Thushara, R.M.; Chandranayaka, S.; Sherman, L.S.; Kemparaju, K.; Girish, K.S. Emerging Roles of hyaluronic acid bioscaffolds in tissue engineering and regenerative medicine. *Int. J. Biol. Macromol.* **2016**, *86*, 917–928. [CrossRef] [PubMed]
11. Research, G.V. Hyaluronic acid market size worth USD 15. Available online: https://www.grandviewresearch.com/press-release/global-hyaluronic-acid-market (accessed on 26 November 2018).
12. Sze, J.H.; Brownlie, J.C.; Love, C.A. Biotechnological production of hyaluronic acid: A mini review. *3 Biotech* **2016**, *6*, 67. [CrossRef] [PubMed]
13. Jing, W.; DeAngelis, P.L. Synchronized chemoenzymatic synthesis of monodisperse hyaluronan polymers. *J. Biol. Chem.* **2004**, *279*, 42345–42349. [CrossRef] [PubMed]
14. Cyphert, J.M.; Trempus, C.S.; Garantziotis, S. Size matters: Molecular weight specificity of hyaluronan effects in cell biology. *Int. J. Cell Biol.* **2015**, *2015*, 1–8. [CrossRef] [PubMed]

15. DeAngelis, P.L. Monodisperse hyaluronan polymers: Synthesis and potential applications. *Curr. Pharm. Biotechnol.* **2008**, *9*, 246–248. [CrossRef] [PubMed]
16. Liu, L.; Liu, Y.; Li, J.; Du, G.; Chen, J. Microbial production of hyaluronic acid: Current state, challenges, and perspectives. *Microb. Cell Fact.* **2011**, *10*, 99. [CrossRef] [PubMed]
17. Jing, W.; DeAngelis, P.L. Dissection of the two transferase activities of the pasteurella multocida hyaluronan synthase: Two active sites exist in one polypeptide. *Glycobiology* **2000**, *10*, 883–889. [CrossRef] [PubMed]
18. DeAngelis, P.L.; Jing, W.; Drake, R.R.; Achyuthan, A.M. Identification and molecular cloning of a unique hyaluronan synthase from pasteurella multocida. *J. Biol. Chem.* **1998**, *273*, 8454–8458. [CrossRef] [PubMed]
19. DeAngelis, P.L. Molecular directionality of polysaccharide polymerization by the *pasteurella multocida* hyaluronan synthase. *J. Biol. Chem.* **1999**, *274*, 26557–26562. [CrossRef] [PubMed]
20. Jing, W. Analysis of the two active sites of the hyaluronan synthase and the chondroitin synthase of *Pasteurella multocida*. *Glycobiology* **2003**, *13*, 661–671. [CrossRef] [PubMed]
21. Kooy, F.K.; Beeftink, H.H.; Eppink, M.H.M.; Tramper, J.; Eggink, G.; Boeriu, C.G. Kinetic and structural analysis of two transferase domains in *Pasteurella multocida* hyaluronan synthase. *J. Mol. Catal. B Enzym.* **2014**, *102*, 138–145. [CrossRef]
22. Kooy, F.K.; Beeftink, H.H.; Eppink, M.H.M.; Tramper, J.; Eggink, G.; Boeriu, C.G. Structural and functional evidence for two separate oligosaccharide binding sites of *Pasteurella multocida* hyaluronan synthase. *Adv. Enzym. Res.* **2013**, *01*, 97–111. [CrossRef]
23. Li, S.; Wang, S.; Fu, X.; Liu, X.W.; Wang, P.G.; Fang, J. Sequential one-pot multienzyme synthesis of hyaluronan and its derivative. *Carbohydr. Polym.* **2017**, *178*, 221–227. [CrossRef] [PubMed]
24. Guo, Y.; Fang, J.; Li, T.; Li, X.; Ma, C.; Wang, X.; Wang, P.G.; Li, L. Comparing substrate specificity of two UDP-sugar pyrophosphorylases and efficient one-pot enzymatic synthesis of UDP-GlcA and UDP-GalA. *Carbohydr. Res.* **2015**, *411*, 1–5. [CrossRef] [PubMed]
25. Kotake, T.; Hojo, S.; Yamaguchi, D.; Aohara, T.; Konishi, T.; Tsumuraya, Y. Properties and physiological functions of UDP-sugar pyrophosphorylase in *Arabidopsis*. *Biosci. Biotechnol. Biochem.* **2007**, *71*, 761–771. [CrossRef] [PubMed]
26. Lau, K.; Thon, V.; Yu, H.; Ding, L.; Chen, Y.; Muthana, M.M.; Wong, D.; Huang, R.; Chen, X. Highly efficient chemoenzymatic synthesis of B1–4-Linked galactosides with promiscuous bacterial B1–4-galactosyltransferases. *Chem. Commun.* **2010**, *46*, 6066. [CrossRef] [PubMed]
27. Litterer, L.A.; Schnurr, J.A.; Plaisance, K.L.; Storey, K.K.; Gronwald, J.W.; Somers, D.A. Characterization and expression of *Arabidopsis* UDP-sugar pyrophosphorylase. *Plant Physiol. Biochem.* **2006**, *44*, 171–180. [CrossRef] [PubMed]
28. Liu, J.; Zou, Y.; Guan, W.; Zhai, Y.; Xue, M.; Jin, L.; Zhao, X.; Dong, J.; Wang, W.; Shen, J.; et al. Biosynthesis of nucleotide sugars by a promiscuous UDP-sugar pyrophosphorylase from *Arabidopsis thaliana* (AtUSP). *Bioorg. Med. Chem. Lett.* **2013**, *23*, 3764–3768. [CrossRef] [PubMed]
29. Muthana, M.M.; Qu, J.; Xue, M.; Klyuchnik, T.; Siu, A.; Li, Y.; Zhang, L.; Yu, H.; Li, L.; Wang, P.G.; et al. Improved one-pot multienzyme (OPME) systems for synthesizing UDP-Uronic acids and glucuronides. *Chem. Commun.* **2015**, *51*, 4595–4598. [CrossRef] [PubMed]
30. Pieslinger, A.M.; Hoepflinger, M.C.; Tenhaken, R. Cloning of glucuronokinase from *Arabidopsis thaliana*, the last missing enzyme of the Myo-Inositol oxygenase pathway to nucleotide sugars. *J. Biol. Chem.* **2010**, *285*, 2902–2910. [CrossRef] [PubMed]
31. Eisele, A.; Zaun, H.; Kuballa, J.; Elling, L. In vitro one-pot enzymatic synthesis of hyaluronic acid from sucrose and *N*-Acetylglucosamine: Optimization of the Enzyme Module System and Nucleotide Sugar Regeneration. *ChemCatChem* **2018**, *10*, 2969–2981. [CrossRef]
32. Nishimoto, M.; Kitaoka, M. Identification of *N*-Acetylhexosamine 1-Kinase in the complete Lacto-*N*-Biose I/Galacto-*N*-Biose Metabolic Pathway in *Bifidobacterium longum*. *Appl. Environ. Microbiol.* **2007**, *73*, 6444–6449. [CrossRef] [PubMed]
33. Chen, Y.; Thon, V.; Li, Y.; Yu, H.; Ding, L.; Lau, K.; Qu, J.; Hie, L.; Chen, X. One-pot three-enzyme synthesis of UDP-GlcNAc Derivatives. *Chem. Commun.* **2011**, *47*, 10815. [CrossRef] [PubMed]
34. Chen, Y.; Li, Y.; Yu, H.; Sugiarto, G.; Thon, V.; Hwang, J.; Ding, L.; Hie, L.; Chen, X. Tailored design and synthesis of heparan sulfate oligosaccharide analogues using sequential one-pot multienzyme systems. *Angew. Chem. Int. Ed.* **2013**, *52*, 11852–11856. [CrossRef] [PubMed]

35. Wahl, C.; Hirtz, D.; Elling, L. Multiplexed capillary electrophoresis as analytical tool for fast optimization of multi-enzyme cascade reactions—Synthesis of nucleotide sugars. *Biotechnol. J.* **2016**, *11*, 1298–1308. [CrossRef] [PubMed]
36. Kotake, T.; Yamaguchi, D.; Ohzono, H.; Hojo, S.; Kaneko, S.; Ishida, H.; Tsumuraya, Y. UDP-sugar pyrophosphorylase with broad substrate specificity toward various monosaccharide 1-Phosphates from pea sprouts. *J. Biol. Chem.* **2004**, *279*, 45728–45736. [CrossRef] [PubMed]
37. Sigma-Aldrich. Uridine 5′-triphosphate trisodium salt dihydrate ≥80% | Sigma-Aldrich. Available online: https://www.sigmaaldrich.com/catalog/product/sigma/94370?lang=de&region=DE&gclid=EAIaIQobChMIzIvUxKPG5AIVx6sYCh1WdAdfEAAYASAAEgKLtvD_BwE (accessed on 10 September 2019).
38. DeAngelis, P.L. Enzymological characterization of the *Pasteurella multocida* hyaluronic acid synthase. *Biochemistry* **1996**, *35*, 9768–9771. [CrossRef] [PubMed]
39. Nunez, H.A.; Barker, R. The metal ion catalyzed decomposition of nucleoside diphosphate sugars. *Biochemistry* **1976**, *15*, 3843–3847. [CrossRef] [PubMed]
40. Jeske, L.; Placzek, S.; Schomburg, I.; Chang, A.; Schomburg, D. BRENDA in 2019: A european ELIXIR core data resource. *Nucleic Acids Res.* **2019**, *47*, D542–D549. [CrossRef] [PubMed]
41. Schomburg, D. BRENDA—Information on EC 2.7.1.43—Glucuronokinase. Available online: https://www.brenda-enzymes.org/enzyme.php?ecno=2.7.1.43#REF (accessed on 11 September 2019).
42. Oria-Hernández, J.; Cabrera, N.; Pérez-Montfort, R.; Ramírez-Silva, L. Pyruvate Kinase Revisited. *J. Biol. Chem.* **2005**, *280*, 37924–37929. [CrossRef] [PubMed]
43. Page, M.J.; Di Cera, E. Role of Na^+ and K^+ in enzyme function. *Physiol. Rev.* **2006**, *86*, 1049–1092. [CrossRef] [PubMed]
44. Li, M.H.; Kwok, F.; Chang, W.R.; Lau, C.K.; Zhang, J.P.; Lo, S.C.L.; Jiang, T.; Liang, D.C. Crystal structure of brain pyridoxal kinase, a novel member of the ribokinase superfamily. *J. Biol. Chem.* **2002**, *277*, 46385–46390. [CrossRef] [PubMed]
45. Lainé-Cessac, P.; Allain, P. Kinetic Studies of the Effects of K^+, Na^+ and Li^+ on the Catalytic Activity of Human Erythrocyte Pyridoxal Kinase. *Enzym. Protein* **1996**, *49*, 291–304. [CrossRef] [PubMed]
46. Jagannath, S.; Ramachandran, K.B. Influence of competing metabolic processes on the molecular weight of hyaluronic acid synthesized by *Streptococcus zooepidemicus*. *Biochem. Eng. J.* **2010**, *48*, 148–158. [CrossRef]
47. Chen, W.Y.; Marcellin, E.; Hung, J.; Nielsen, L.K. Hyaluronan molecular weight is controlled by UDP-*N*-Acetylglucosamine concentration in *Streptococcus zooepidemicus*. *J. Biol. Chem.* **2009**, *284*, 18007–18014. [CrossRef] [PubMed]
48. Sheng, J.Z.; Ling, P.X.; Zhu, X.Q.; Guo, X.P.; Zhang, T.M.; He, Y.L.; Wang, F.S. Use of induction promoters to regulate hyaluronan synthase and UDP-Glucose-6-Dehydrogenase of *Streptococcus zooepidemicus* expression in *Lactococcus lactis*: A case study of the regulation mechanism of hyaluronic acid polymer. *J. Appl. Microbiol.* **2009**, *107*, 136–144. [CrossRef] [PubMed]
49. DeAngelis, P.L.; Papaconstantinou, J.; Weigel, P.H. Isolation of a *Streptococcus pyogenes* gene locus that directs hyaluronan biosynthesis in acapsular mutants and in heterologous bacteria. *J. Biol. Chem.* **1993**, *268*, 14568–14571. [PubMed]
50. Weigel, P.H.; DeAngelis, P.L. Hyaluronan synthases: A decade-plus of novel glycosyltransferases. *J. Biol. Chem.* **2007**, *282*, 36777–36781. [CrossRef] [PubMed]
51. Blank, L.M.; Hugenholtz, P.; Nielsen, L.K. Evolution of the hyaluronic acid synthesis (*has*) operon in *Streptococcus zooepidemicus* and other pathogenic *Streptococci*. *J. Mol. Evol.* **2008**, *67*, 13–22. [CrossRef] [PubMed]
52. Williams, K.J.; Halkes, K.M.; Kamerling, J.P.; DeAngelis, P.L. Critical elements of oligosaccharide acceptor substrates for the *Pasteurella multocida* hyaluronan synthase. *J. Biol. Chem.* **2006**, *281*, 5391–5397. [CrossRef] [PubMed]
53. Boeriu, C.G.; Springer, J.; Kooy, F.K.; van den Broek, L.A.M.; Eggink, G. Production methods for hyaluronan. *Int. J. Carbohydr. Chem.* **2013**, *2013*, 1–14. [CrossRef]
54. Chong, B.F.; Blank, L.M.; Mclaughlin, R.; Nielsen, L.K. Microbial hyaluronic acid production. *Appl. Microbiol. Biotechnol.* **2005**, *66*, 341–351. [CrossRef] [PubMed]
55. De Oliveira, J.D.; Carvalho, L.S.; Gomes, A.M.V.; Queiroz, L.R.; Magalhães, B.S.; Parachin, N.S. Genetic basis for hyper production of hyaluronic acid in natural and engineered microorganisms. *Microb. Cell Fact.* **2016**, *15*, 119. [CrossRef] [PubMed]

56. Iyer, P.V.; Ananthanarayan, L. Enzyme stability and stabilization—aqueous and non-aqueous environment. *Process Biochem.* **2008**, *43*, 1019–1032. [CrossRef]
57. Mandawe, J.; Infanzon, B.; Eisele, A.; Zaun, H.; Kuballa, J.; Davari, M.D.; Jakob, F.; Elling, L.; Schwaneberg, U. Directed evolution of hyaluronic acid synthase from *Pasteurella multocida* towards high-molecular-weight hyaluronic acid. *ChemBioChem* **2018**, *19*, 1414–1423. [CrossRef] [PubMed]
58. Lee, H.G.; Cowman, M.K. An agarose gel electrophoretic method for analysis of hyaluronan molecular weight distribution. *Anal. Biochem.* **1994**, *219*, 278–287. [CrossRef] [PubMed]
59. Di Ferrante, N. Turbidimetric measurement of acid mucopolysaccharides and hyaluronidase activity. *J. Biol. Chem.* **1956**, *220*, 303–306. [PubMed]
60. Chen, Y.-H.; Wang, Q. Establishment of CTAB turbidimetric method to determine hyaluronic acid content in fermentation broth. *Carbohydr. Polym.* **2009**, *78*, 178–181. [CrossRef]
61. Oueslati, N.; Leblanc, P.; Harscoat-Schiavo, C.; Rondags, E.; Meunier, S.; Kapel, R.; Marc, I. CTAB Turbidimetric method for assaying hyaluronic acid in complex environments and under cross-linked form. *Carbohydr. Polym.* **2014**, *112*, 102–108. [CrossRef] [PubMed]
62. Andexer, J.N.; Richter, M. Emerging enzymes for ATP regeneration in biocatalytic processes. *ChemBioChem* **2015**, *16*, 380–386. [CrossRef] [PubMed]
63. De Luca, C.; Lansing, M.; Martini, I.; Crescenzi, F.; Shen, G.J.; O'Regan, M.; Wong, C.H. Enzymic synthesis of hyaluronic acid with regeneration of sugar nucleotides. *J. Am. Chem. Soc.* **1995**, *117*, 5869–5870. [CrossRef]

International Journal of
Molecular Sciences

MDPI

Article

Bioproduction of Quercetin and Rutinose Catalyzed by Rutinosidase: Novel Concept of "Solid State Biocatalysis"

Jana Kapešová [1], Lucie Petrásková [1], Kristína Markošová [2], Martin Rebroš [2], Michael Kotik [1], Pavla Bojarová [1] and Vladimír Křen [1,*]

[1] Institute of Microbiology of the Czech Academy of Sciences, Laboratory of Biotransformation, Vídeňská 1083, CZ 14220 Prague 4, Czech Republic; janakapesova@gmail.com (J.K.); petraskova@biomed.cas.cz (L.P.); kotik@biomed.cas.cz (M.K.); bojarova@biomed.cas.cz (P.B.)

[2] Institute of Biotechnology, Slovak University of Technology, Radlinského 9, 81237 Bratislava, Slovakia; kristina.markosova@stuba.sk (K.M.); martin.rebros@stuba.sk (M.R.)

* Correspondence: kren@biomed.cas.cz; Tel.: +420-296-442-510

Received: 7 February 2019; Accepted: 28 February 2019; Published: 5 March 2019

Abstract: Quercetin is a flavonoid largely employed as a phytochemical remedy and a food or dietary supplement. We present here a novel biocatalytic methodology for the preparation of quercetin from plant-derived rutin, with both substrate and product being in mostly an undissolved state during biotransformation. This "solid-state" enzymatic conversion uses a crude enzyme preparation of recombinant rutinosidase from *Aspergillus niger* yielding quercetin, which precipitates from virtually insoluble rutin. The process is easily scalable and exhibits an extremely high space-time yield. The procedure has been shown to be robust and was successfully tested with rutin concentrations of up to 300 g/L (ca 0.5 M) at various scales. Using this procedure, pure quercetin is easily obtained by mere filtration of the reaction mixture, followed by washing and drying of the filter cake. Neither co-solvents nor toxic chemicals are used, thus the process can be considered environmentally friendly and the product of "bio-quality." Moreover, rare disaccharide rutinose is obtained from the filtrate at a preparatory scale as a valuable side product. These results demonstrate for the first time the efficiency of the "Solid-State-Catalysis" concept, which is applicable virtually for any biotransformation involving substrates and products of low water solubility.

Keywords: *Aspergillus niger*; quercetin; rutin; rutinose; rutinosidase; "solid-state biocatalysis"

1. Introduction

Flavonoids, which occur in plants, vegetables and fruits, are generally considered as efficient antioxidants and chemoprotectants. They are a plentiful but a structurally homogeneous group of plant natural products, whose structure is derived from phenylchromene hydroxylated or methoxylated at various positions. A number of flavonoids are present in natural sources in the form of glycosides. The flavonol quercetin is one of the most common and abundant flavonoids in the plant kingdom. It is a prominent bioactive flavonol in the human diet and its daily intake has increased considerably due to its use as a food supplement [1] and mainly due to the general medical recommendation to "eat five servings of fruits and vegetables a day" [2].

Flavonol quercetin (3,2-(3,4-dihydroxyphenyl)-3,5,7-trihydroxy-4*H*-1-benzopyran-4-one; **1**) frequently occurs in the human diet (typically in apples, onions, buckwheat etc.) in the form of glycosides—the most prominent being quercetin-3-*O*-β-D-glucopyranoside (isoquercitrin) [3] and quercetin-3-*O*-rutinoside (rutin; **3**) (Figure 1) [4]. They have a broad range of positive effects in humans [5] and they are used in a number of medicines and over-the-counter preparations and often in food supplements. Quercetin is largely utilized as a phytochemical remedy for a variety of diseases [5], such as diabetes/obesity and circulatory dysfunctions, inflammation and mood disorders or as an adaptogen. Furthermore, it may be used as an efficient and affordable radioprotectant (e.g., for a large population in case of attack with a "dirty bomb"—radiological dispersal device that combines radioactive material with conventional explosives) [6]. Quercetin used to be considered potentially harmful, since it gave a positive reaction in the Ames' test, which implied its mutagenicity and potential cancerogenicity [7]. Nevertheless, these doubts have recently been dispelled and the Federal Drug Agency in the US issued GRAS notice for pure quercetin [8]. This certification led to the recent commercial boom in quercetin marketing.

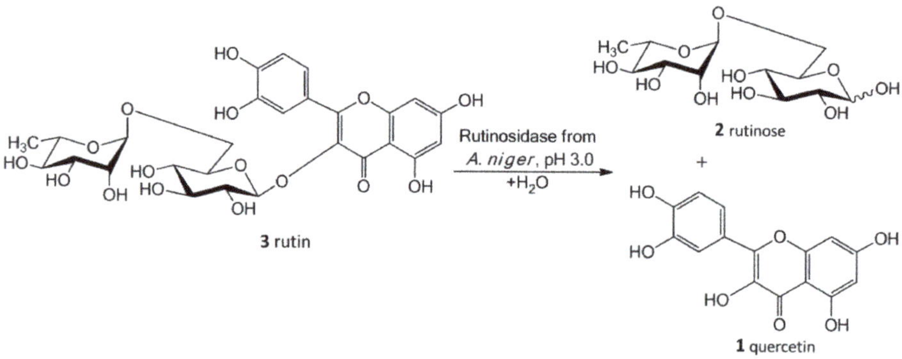

Figure 1. Bioconversion of rutin to quercetin using rutinosidase.

The structure of quercetin is quite unique in terms of antioxidant and antiradical activity: the catechol moiety together with the free OH group at C-3 in conjugation with the C-4 oxo group ensures electron delocalization on ring B. Furthermore, the configuration of OH groups at C-3, -5 and -7 together with the C-4 oxo group make quercetin one of the most potent antioxidants ever discovered [5]. The free C-3 OH group, often glycosylated in natural quercetin glycosides (e.g., rutin, quercitrin, isoquercitrin), is crucial for the high antioxidant activity of quercetin [9,10]. This is the reason why free quercetin has a considerably higher antioxidant activity than its glycosides.

The direct extraction of quercetin from natural sources is impractical, mostly due to the complexity of the matrix and its occurrence in various (glycosylated) forms. An alternative method described in the literature is the oxidation of taxifolin (dihydroquercetin), obtained from Siberian woods, with $NaNO_2$ [11]. This multistep method does not yield high-purity quercetin and uses relatively toxic chemicals.

A common starting material for quercetin production is rutin (**3**)—a commodity chemical obtained from various plant materials by extraction (typically from the Brazilian tree Fava d'anta, *Dimorphandra mollis*). Quercetin is currently prepared by acid hydrolysis of rutin, for example, by boiling with diluted HCl [12,13]. This process yields quercetin and the monosaccharides D-glucose and L-rhamnose because rutinose bound to quercetin via the β-glycosidic linkage is completely hydrolyzed. Boiling in a strong acid may lead to a partial decomposition of all reactants, which deteriorates the purity and quality of the final product. Such a product can also hardly be declared to be of "bio" quality.

There are also enzymatic methods for producing quercetin from rutin [13]. These methods typically employ mixtures of enzymes (e.g., β-D-glucosidases and α-L-rhamnosidases) that yield mixtures of the respective monosaccharides and quercetin. Since rutin has a relatively low water solubility (5 g/L at pH 5; 15 g/L at pH 8) and all enzymatic methods described so far require a completely dissolved rutin, they employ cosolvents such as MeOH, EtOH or DMSO [14,15]. The use of cosolvents complicates the procedure as they are flammable, toxic and their residues may contaminate the final product (especially DMSO, which is not easy to remove by evaporation). Moreover, co-solvents often inactivate the enzymes employed and thus decrease the final yields and extend reaction times. Therefore, their use must always be a compromise between the increase in rutin solubility and enzyme denaturation. The low rutin solubility also limits the possibilities of enzyme recycling or the use of immobilized enzymes-cf. [16].

Our new biosynthetic method allows working with rutin in highly concentrated aqueous suspensions (up to ca 300 g/L, ca 0.5 M) without any co-solvents. This method directly yields a precipitated product (**1**) in a virtually quantitative yield that can easily be filtrated and purified just by washing at the filter. The filtrate contains a still active enzyme, which can be reused despite the presence of rutinose from the first reaction—rutinose has no inhibitory activity on the enzyme. The rare and so far unexplored sugar rutinose (**2**) is produced as a side product and can be crystallized from the concentrated filtrate in a high purity (Figure 1). This disaccharide has a potential application in cosmetics. We present here a novel concept of "Solid State Biocatalysis" that enables an exceptionally high space-time-yield.

2. Results

2.1. Production of Rutinosidase

2.1.1. Production of the Wild-Type Rutinosidase from *Aspergillus niger*

The production and purification of the wild-type rutinosidase from *A. niger* have been previously described in detail [17]. In the scope of this work, the use of the crude, wild-type enzyme for the production of quercetin and rutinose has been tested. Besides the production of rutinosidase (day 6; rutinosidase activity 0.31 U/mL), the co-production of α-L-rhamnosidase (0.30 U/mL) and β-D-glucosidase (0.49 U/mL) during fermentation were observed. This made the use of the crude enzyme unfeasible, mainly due to the parallel cleavage of rutinose and other unwanted reactions. Moreover, the crude wild-type enzyme is not stable (probably due to the presence of proteases in the medium), which was also demonstrated in the tests with rutin bioconversion (see Section 2.2.1). In summary, without enzyme purification, which is not feasible economically at a large scale, the crude wild-type enzyme cannot be used for this process.

2.1.2. Production of Recombinant Rutinosidase from *Aspergillus niger* in *Pichia pastoris*

After cloning procedure we observed diminished stability of crude recombinant rutinosidase below pH 3.5 in the cultivation media that were acidified during cultivation in flasks. This instability was obviously caused by the production of (unspecified) serine protease(s), because addition of the protease inhibitor phenylmethanesulfonylfluoride (PMSF, 1 mM) fully restored enzyme stability [18]. The stability of the enzyme was evaluated in the deglycosylation reaction of rutin (100 g/L). We tested flask production in the BMMH medium with the pH set to values of 3.0; 4.0; 5.0; 6.0. Rutinosidase activity in the medium was 0.080; 0.056; 0.062; and 0.046 U/mL, respectively. Although the specific enzyme production was the highest in the most acidic medium (pH 3.0), the performance of the enzyme produced under these conditions was the worst since the conversion was not completed even after 24 h. The highest initial activity and the highest stability were found for the enzyme produced at pH 5.0 (Figure 2). This figure shows the fastest turnover of the substrate with enzyme obtained by the cultivation at pH 5.0 (green curve), which is the most stable. Enzyme obtained from the medium of pH 3.5 has a high activity but its overall performance is afflicted by its lower stability. This also demonstrates the importance of pH control of the fermentation for enzyme stability.

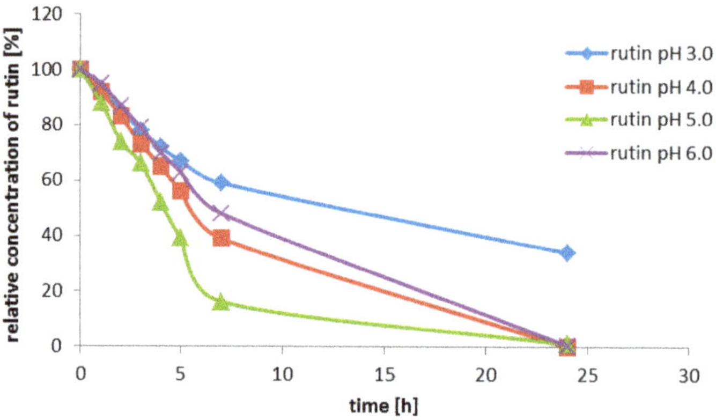

Figure 2. Bioconversion of rutin to quercetin using crude recombinant rutinosidase produced at various pH (rutin concentration 100 g/L; reaction volume 10 mL; 40 °C; pH 3); rutinosidase was produced in the media with the starting pH set to 3–6.

2.1.3. Scale-Up of the Production of Recombinant Rutinosidase

The production of recombinant rutinosidase by *P. pastoris* KM71H was scaled up in 3-L laboratory bioreactors. Glycerol was depleted after approximately 20 h, during which the exponential growth of biomass was observed (Figure 3). Then, two methanol pulses of 3 g/L were added at hour 21 and hour 32. Fed-batch methanol feeding was started at hour 35, according to the optimized protocol described by [19]. During the fed-batch phase, methanol was fed according to the actual level of dissolved oxygen and never exceeded a concentration of 3 g/L, which is toxic for this *P. pastoris* Mut^S strain. The concentration of methanol started to increase slightly at the end of the fermentation and also the growth of biomass steadied, which indicated that the biomass reached the stationary phase and the fermentation was terminated. SDS-PAGE electrophoresis (Figure S1, Supplementary Material) showed a band of approx. 66 kDa, representing the produced rutinosidase. The absence of other significant protein bands demonstrates the great advantage of controlled fermentation, which produces no other extracellular proteins than the desired rutinosidase. The overall productivity of the produced rutinosidase was 5.69 $mg_{prot} \cdot L^{-1} \cdot h^{-1}$.

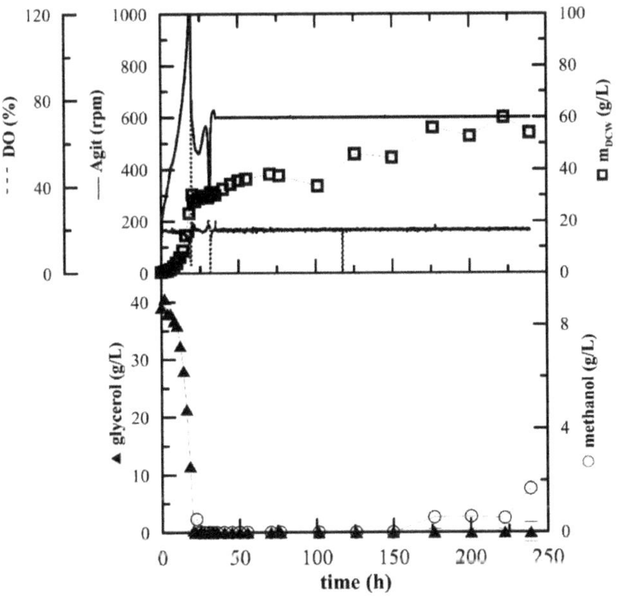

Figure 3. Fed-batch cultivation of *P. pastoris* expressing rutinosidase with methanol feeding according to the actual level of dissolved oxygen. Conditions: 1.5 L BSM, 5% inoculum, 30 °C, pH 5.0, DO 20%, stirring cascade 50–1000 rpm; after 35 h fixed at 600 rpm.

2.2. Bioconversion of Rutin to Quercetin

2.2.1. Bioconversion of Rutin to Quercetin Using Wild-Type Rutinosidase

The first tests of a larger-scale conversion of rutin to quercetin were performed with the wild-type enzyme. The reaction was successfully accomplished at analytical scale with the purified wild-type rutinosidase (up to 100 g/L rutin). However, for a practical large-scale biotechnological application protein purification is not economically feasible. Therefore we also tested crude rutinosidase (centrifuged medium from the cultivation of *A. niger*, adjusted to pH 3 with phosphoric acid, diluted twice with water). The rutin concentrations were 50, 100, 150 and 200 g/L. Only the reaction with 50 g/L was completed within 24 h whereas higher concentrations of rutin led to incomplete conversion (Figure S2, Supplementary Material). It seems that the enzyme is partially inactivated during the reaction with a higher rutin concentration, possibly by some proteases in the crude medium. Moreover, the crude medium also contains a considerable activity of α-L-rhamnosidase (rutinosidase 0.305 U/mL, α-L-rhamnosidase 0.30 U/mL), which causes degradation of rutinose—valuable by-product of this bioconversion.

2.2.2. Bioconversion of Rutin to Quercetin Using Recombinant Rutinosidase

Optimization of Reaction Conditions

The recombinant rutinosidase from *A. niger* produced in *P. pastoris* has a temperature optimum of 50 °C at pH 3.5; temperatures exceeding 55 °C resulted in a loss of activity [17]. The stability of the recombinant (crude, dialyzed) enzyme at different pH and temperatures was tested. The enzyme loses virtually all activity in ca 2 h at 50 °C. (Figure S3a, Supplementary Data). Therefore a lower temperature (40 °C), where the enzyme has an activity of ca 60% compared to the optimum temperature but is quite stable, was selected.

Nevertheless, even at 50 °C the performance of rutinosidase during rutin bioconversion was very satisfactory. The complete conversion of rutin at a concentration of 200 g/L was accomplished within 5–6 h (Figure S3b, Supplementary Material). It is obvious that the enzyme in the presence of substrate is much more stable than in a mere buffer.

The pH optimum of recombinant rutinosidase is at pH 3.0. At pH 5.0 the enzyme still maintains 50% of its maximum activity and at pH 2.0 and pH 7.0 it is virtually inactive [17]. To support the above hypothesis the performance of the enzyme in a real reaction system with a high rutin concentration (200 g/L) was tested at different pH. The results showed (Figure 4b) that at pH 2.0 the enzyme is inactive and its performance is suboptimal at pH 5.0. However, from pH 2.5 to 4.5 the enzyme performance is—in this experimental setup—practically the same. Therefore, for further experiments pH 3.0 was selected as the optimum one for rutin bioconversion.

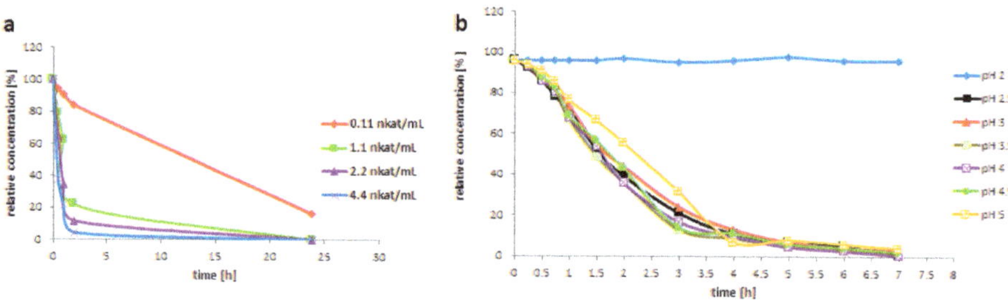

Figure 4. Optimization of reaction conditions for bioconversion of rutin to quercetin. (**a**) Bioconversion of rutin using various amounts of pure rutinosidase (rutin 200 g/L; reaction volume 3 mL; 35 °C, pH 3.0). (**b**) Bioconversion of rutin in 0.2 M glycine buffer at various pH (rutin 200 g/L, reaction volume 3 mL, 35 °C, purified rutinosidase 0.2 U/mL).

Furthermore, the activity of enzyme in the reaction system containing 200 g rutin/L was optimized. The critical rutinosidase activity seems to be ca 0.1 U/mL. At a rutinosidase activity of 0.2 U/mL, the reaction was completed within ca 6 h (Figure 4a).

Enzyme Storage and Stability

For the purposes of potential biotechnological applications, it is vital to assess the possibilities of long-term storage of the enzyme. The enzyme (both wild type and purified recombinant) is stable for several months at 6 °C. Moreover, we found that recombinant rutinosidase may be lyophilized from the crude medium without any notable loss of catalytic activity. The rutinosidase activity of the lyophilized sample resuspended in 50 mM citrate-phosphate buffer pH 3.0 was at least 90% compared to the fresh enzyme with *p*NP-rutinoside. The comparison of rutin bioconversion reactions with the lyophilized enzyme after shock or slow freezing (Figure S4a,b Supplementary Material) shows that shock freezing is a preferable method due to its mildness and full preservation of rutinosidase activity.

Optimization of the Substrate Concentration and Scale-Up

The substrate concentration is crucial to achieve the optimal space-time-yield. The substrate concentrations considerably higher than 100 g/L, for example, 200, 300, 450 g/L rutin in reaction volumes of 30 and 100 mL and also a concentration of 450 g/L in a volume of 500 mL were tested. The crude recombinant enzyme prepared in the fermenter under controlled pH (0.8 U/mL) gave 100% conversion in all reaction setups within 24 h. For practical applications, 200 g/L (Figure 5) seems to be the highest suitable substrate concentration due to the fact that at very high rutin concentrations the reaction mixture tends to become significantly more viscous (especially after reaction completion) and it needs to be diluted with water for future processing.

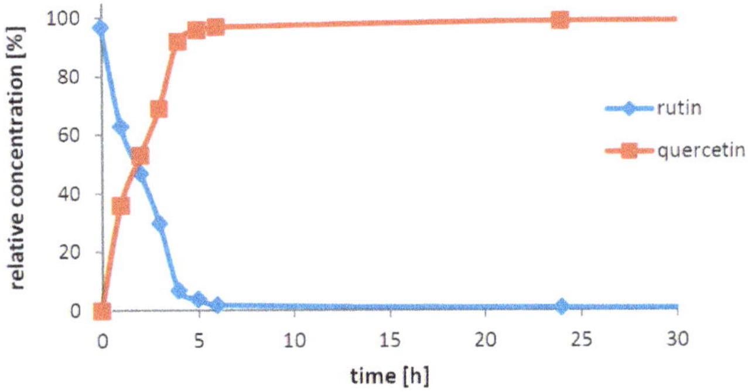

Figure 5. Bioconversion of rutin to quercetin under optimum conditions. Rutin 200 g/L; reaction volume 30 mL; 40 °C; pH 3.0; crude recombinant rutinosidase activity 0.8 U/mL.

Reaction scale-up was tested with a substrate concentration of 200 g/L in volumes of 50, 100, 250 mL and 2 L (Figure S5, Supplementary Material) using as little as 0.2 U/mL rutinosidase. In all the reaction set-ups, the conversion was completed within ca 6 h yielding quercetin in virtually quantitative yield.

2.3. The Concept of "Solid-State Biocatalysis"

Since the bioconversion of rutin into quercetin and rutinose proceeds in a very thick suspension, we were interested to see how the reaction runs at the microscopic level. The monitoring of the reaction mixture with an optical microscope (Figure 6) showed that the starting material was indeed present in crystalline form and over the course of the reaction (2 and 5 h) multiple small crystals of the product (most probably quercetin) were formed, which gradually substituted the large crystals of rutin as the reaction proceeded (Figure 6c).

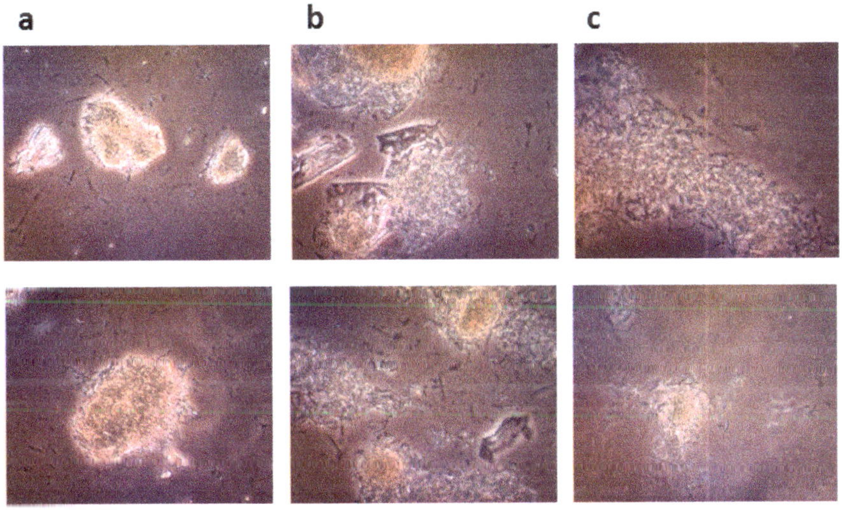

Figure 6. Microscopic monitoring of "Solid-State Biocatalysis" after (**a**) 0 h, (**b**), 2 h and (**c**), 5 h of reaction (rutin 100 g/L, crude recombinant rutinosidase). OLYMPUS CX41 optical microscope equipped with OLYMPUS U-CMAD3 digital camera, magnification 400×.

3. Discussion

The above results clearly demonstrate that the rutinosidase from *Aspergillus niger* is a highly attractive enzyme for biotechnological application. Its heterologous expression elegantly solved the problem of contaminating α-L-rhamnosidase and β-D-glucosidase activities co-produced with the wild-type enzyme, and, at the same time, allowed a multiple-litre production scale-up, which can be further upscaled if convenient. Moreover, due to the production selectivity, the robust crude medium can directly be applied for a one-step bioproduction of a GRAS-classified phytochemical quercetin with a high space-time yield. Here, we must highlight the necessity of a thorough search for the optimum reaction conditions. As previously confirmed with other glycosidases, the temperature optimum may not copy the temperature stability [20]. Admittedly, the mere activity measurement profiles under given conditions do not always reflect the real situation during the biocatalytical process. There, the enzyme active site is protected against the detrimental effect of extreme conditions such as high temperature by the proceeding biocatalytical reaction. This was clearly visible comparing the results of stability measurements and the real reaction outcome at 50 °C (Section 2.2.2). Interestingly, this effect is observed even though no inhibition by the released hydrolytic product (rutinose) occurred. Rutinosidase inhibition with rutinose has been tested and up to 0.5 M rutinose (corresponding to ca 300 g/L starting concentration of rutin) the rutinosidase activity remains unchanged (data not shown). Apparently, since the enzyme active site is constantly occupied by the incoming reaction substrate in excess, its architecture is well protected against the impact of detrimental conditions. Thus, the enzyme is obviously more stable in a bioconversion reaction with a high rutin concentration than the enzyme incubated under the same conditions in a mere buffer.

The crucial finding presented in this work is the particular biocatalytic set-up that we denoted as "Solid-State-Biocatalysis." From the above data, it is obvious that the bioconversion of rutin into quercetin and rutinose catalyzed by rutinosidase proceeds in a very thick suspension (tested up to 450 g/L ≈ 0.7 M), where both the starting material and product remain primarily in the solid phase (the typical solubility of both reactants is ca 1–3 g/L, with quercetin being less soluble than rutin). Numerous authors tested this and other similar reactions, typically the partial hydrolysis of rutin into isoquercitrin with α-L-rhamnosidase, with the addition of various cosolvents [21]. In these reaction set-ups, authors barely achieved concentrations of dissolved rutin higher than approx. 10 g/L, following the commonly accepted scheme that enzymatic reactions proceed solely or at least in majority in the liquid phase. In the present case, in contrast, the majority of substrate is present in the solid phase. We suppose that the reaction proceeds in the (over)saturated microenvironment solution of the substrate crystals. The formed product, also badly water soluble, precipitates quickly out of this microenvironment, thereby constantly shifting the reaction equilibrium to the full conversion.

There are obviously two major driving forces of the reaction: (i) the high affinity of the biocatalyst for the substrate, enabling its conversion at a low concentration and (ii) the thermodynamic shift of equilibrium towards product formation caused by continuous product "removal" due to its precipitation. This type of reaction apparently runs well when both reactants are poorly soluble and the catalyst is efficient. Interestingly, when a cosolvent (e.g., 5% DMSO) is added into the reaction mixture, the reaction is slower and the conversion is not complete. This corroborates the above hypothesis because with the addition of the cosolvent, the product occurs in higher concentrations (inhibition with a higher concentration of quercetin) in the reaction mixture due to its increased solubility. A partial inactivation of the enzyme by the cosolvent cannot be excluded either.

Another advantage of the "Solid-State Biocatalysis" setup is the resistance of rutinosidase towards product inhibition by the disaccharide rutinose. On the contrary, the accumulation of rutinose in the reaction mixture and the running catalytic reaction stabilize the enzyme, as observed in our stability studies (see Section 2.2.2 and Figure S3, Supplementary Material). Rutinose in a higher concentration may also diminish the solubility of quercetin, lowering thus its concentration.

The main product of this enzymatic process is pure quercetin that does not involve any harmful and irritant chemicals, which makes it applicable in the cosmetics and food industries as a "Bio-quality" product. This procedure also yields the valuable side product rutinose, which has not been available in multigram/kilogram scale for a reasonable price so far. Rutinose can be obtained from the filtrate of the reaction mixture after short boiling with charcoal, calcium hydroxide and Celite (each 0.5 g/L), which removes protein and phosphate impurities and decolorizes the solution. After filtration and evaporation, pure rutinose (>97%) is obtained by crystallization or lyophilization. This opens possibilities for research and potential application of this disaccharide for example, in cosmetic applications because α-L-rhamnose and/or α-L-rhamnosides were found to interact with specific receptors on keratinocytes, which play an important role in cell and (skin) tissue aging [22].

4. Materials and Methods

4.1. Materials

The enzymes and buffers for DNA manipulations were obtained from New England Biolabs (Ipswich, MA, USA). Media components were from Oxoid (Basingstoke, UK) or Carl-ROTH (Karlsruhe, Germany). Rutin was from Alchimica (Prague, Czech Republic) and other chemicals were purchased from Sigma-Aldrich.

4.2. Biological Material

An EasySelect *Pichia pastoris* KM71H Expression Kit was purchased from Invitrogen (Waltham, MA, USA). This expression system employs the AOX1 (alcohol oxidase 1) promoter, inducible by methanol.

The culture of *Aspergillus niger* K2 CCIM is deposited in the Collection of Microorganisms of the Institute of Microbiology of the Czech Academy of Sciences, Prague. The culture is maintained on slants [g/L]: agar-agar, 20; bacto-peptone, 5; and malt extract, 35.

4.3. Media

The inoculum for *P. pastoris* cultivation was prepared in BMGY medium (Buffered Glycerol-Complex Medium) [g/L]: yeast extract, 10; peptone, 20; 100 mM potassium phosphate, pH 6.0; Yeast Nitrogen Base (YNB, Oxoid cz., Brno, Czech Republic), 13.4; biotin 0.0004; glycerol, 10.

Buffered Methanol-Complex Medium (BMMY) has the same composition as BMGY but instead of 1% (*w*/*v*) glycerol methanol (0.5%, *v*/*v*) is added.

For large-scale productions, minimal media were used. BMGH medium (Buffered Minimal Glycerol Medium) [g/L]: 100 mM potassium phosphate pH 6.0; YNB, 13.4; biotin 0.0004; glycerol, 10 for overnight preculture and BMMH medium (Buffered Minimal Methanol Medium) [g/L]: 100 mM potassium phosphate pH 6.0; YNB, 13.4; biotin 0.0004; methanol, 5).

Fed-batch fermentations were carried out in BSM medium (Basal Salt Medium) [g/L]: 85% H_3PO_4, 26.7 mL; $CaSO_4{\cdot}2H_2O$, 1.17; K_2SO_4, 18.2; $MgSO_4{\cdot}7H_2O$, 14.9; KOH, 4.13; and glycerol, 40; and supplemented with 4.35 mL/L of PTM_1 (trace salts solution) [g/L]: $CuSO_4{\cdot}5H_2O$, 6; NaI, 0.08; $MnSO_4{\cdot}H_2O$, 3; $Na_2MoO_4{\cdot}2H_2O$, 0.2; H_3BO_3, 0.02; $CoCl_2$, 0.5; $ZnCl_2$, 20; $FeSO_4{\cdot}7H_2O$, 65; biotin, 0.2; H_2SO_4 conc. 9.2 g). Methanol added in fed-batch experiments was also supplemented with PTM_1 (1.2 mL/L pure methanol).

The production medium for *A. niger* cultivation of consisted of [g/L]: rutin, 5.0; KH_2PO_4, 15.0; NH_4Cl, 4.0; KCl, 0.5; yeast extract, 5.0; casein hydrolysate, 1.0; and 1.0 mL of trace element Vishniac solution [23] at pH 5.0. The pH of the medium was adjusted to 5.0. After sterilization, each flask was supplemented with 1.0 mL of sterile 10% $MgSO_4{\cdot}7H_2O$ (*w*/*v*).

4.4. Preparation of Enzymes

4.4.1. Cultivation of *Aspergillus niger*

The production medium (200 mL) was inoculated with a suspension of spores in a 0.1% (*v*/*v*) Tween 80 solution and cultivated in 500-mL Erlenmeyer flasks on a rotary shaker at 28 °C and 250 rpm.

The mycelium of *A. niger* K2 CCIM was cultivated for 12 days at 28 °C and 250 rpm in Erlenmeyer flasks in the presence of 0.5% *w*/*v* rutin as the rutinosidase inducer. The mycelium was subsequently removed by filtration through an asbestos-cellulose filter (C10, Vertex CZ, Prague, Czech Republic). The filtrate was directly used (as a crude wild rutinosidase) for rutin conversion.

4.4.2. Heterologous Expression of Rutinosidase in *Pichia pastoris*

The expression vector pPICZαA-RUT, obtained as described previously [17] was linearized with restriction endonuclease *Sac*I and electroporated to competent *P. pastoris* cells according to the manufacturer's instructions (EasySelect *Pichia* Expression Kit, Invitrogen; Waltham, MA, USA). The electroporated cells were grown at various concentrations of zeocin (100 μg/mL) on YPD (Yeast Extract Peptone Dextrose Medium; yeast extract OXOID 1%, bacteriological peptone OXOID 2%, glucose 2% (LACHNER, Neratovice, Czech Republic)) agar plates for 2 days at 28 °C.

The production of recombinant rutinosidase was performed according to the manufacturer's instructions as follows: the colonies were inoculated into 100 mL of BMGY medium pH 6.0 and incubated overnight with shaking at 28 °C. The cells were collected by centrifugation (5000× *g*, 10 min, 20 °C) and the pellet was resuspended in 30 mL of BMMY medium (see Section 4.3 Media) in a 300 mL baffled flask. The production of rutinosidase was induced by the addition of methanol (0.5% *v*/*v*) every 24 h for 4 days. The flasks were incubated at 28 °C and 220 rpm.

For the large-scale production of rutinosidase, we used 1 L of BMGH medium for overnight preculture. The cells were then collected and resuspended in 200 mL of BMMH medium and incubated at 28 °C on a rotary shaker with methanol induction (0.5% *v*/*v*) every 24 h for 4 days.

4.4.3. Fermentation Scale-Up

Fermentation Media

The inoculum was prepared in BMGY medium. Fed-batch fermentations were carried out in BSM medium; methanol added in fed-batch experiments was also supplemented with PTM_1 (1.2 mL/L methanol).

Bioreactor Cultivation

The inoculum for the fermentation cultivation was prepared in 100 mL of BMGY medium. The fermentation was essentially performed as described previously [20] in 3-L laboratory fermenters (Brunswick BioFlo® 115, Eppendorf, Hamburg, Germany). Then, 1.5 L of BSM media supplemented with 6.53 mL of PTM_1 was inoculated with inoculum (OD_{600} approx. 10–12) to a concentration of 5% *v*/*v*. The fermentation conditions were as follows: 30 °C, pH 5 maintained with ammonium solution (28–30%), DO (dissolved oxygen) 20% maintained by agitation cascade from 50 to 1000 rpm, aeration 0.66 vvm with the addition of 200 μL of Struktol J650 (Schill + Seilacher "Struktol" GmbH, Hamburg, Germany) as an antifoaming agent. After the complete depletion of glycerol (approx. 20 h), two methanol pulses of 3 g/L were added after 20 and 31 h. After the depletion of the second pulse, the agitation was fixed at 600 rpm, the agitation cascade was stopped and additional methanol (3 g/L) was added. The methanol feed was connected to the actual level of dissolved oxygen as described in Reference [20]. Whenever the level of DO rose above 20%, the methanol feed was turned on by an automated program and when the DO level rose above 30%, signalling an excess of methanol and the inability of the culture to utilize it, the pump was turned off again.

4.5. *Purification of Recombinant Rutinosidase*

Recombinant rutinosidase was purified from the culture medium of *P. pastoris* after 6 days of cultivation with methanol induction. The cells were harvested by centrifugation (5000× *g*, 10 min at 4 °C). The supernatant was dialyzed against 6 L of 10 mM sodium acetate buffer, pH 3.6, for 2 h (dialysis tubing cellulose membrane, Sigma-Aldrich, cut-off 10 kDa). The pH of the solution was then adjusted to 3.6 with 10% acetic acid and filtered. This solution was loaded into a Fractogel EMD SO_3^- column (15 × 100 mm) in 10 mM sodium acetate buffer, pH 3.6. The protein was eluted using a linear gradient of 0–1 M NaCl (5 mL/min). Fractions were collected and then analyzed for rutinosidase activity using *p*-nitrophenol rutinoside as a substrate. The fractions containing rutinosidase activity were concentrated by ultrafiltration using cellulose membranes with a 10 kDa cut-off (Millipore, Merck, Darmstadt, Germany). The concentrated protein was then purified to homogeneity by gel filtration in a Superdex 200 10/300 GL column (10 × 300 mm, 10 mM citrate-phosphate buffer, pH 5.0, 150 mM NaCl).

Protein concentrations were determined by Bradford assay calibrated for bovine serum albumin. The purity of recombinant rutinosidase was checked by 12% SDS-PAGE.

4.6. Storage of Enzymes

The purified or crude rutinosidase (as a cultivation medium from transformed *P. pastoris*) could be stored with no notable loss of activity in a fridge (ca 6 °C) for a minimum of 6 months. For lyophilization, the crude medium was rapidly frozen in a liquid nitrogen bath (shock freezing) or frozen in a deep freezer (−80 °C, slow freezing). The material was then lyophilized to dryness. The loss of specific activity of rutinosidase was in both cases under 10%. The lyophilized enzyme can be stored in a fridge in a tightly closed vessel for a minimum of 6 months with no loss of activity.

4.7. Enzyme Activity Assay

The rutinosidase activity was measured spectrophotometrically using *p*-nitrophenyl rutinoside as a substrate at a starting concentration of 2 mM. The reaction mixture contained substrate (10 mM solution, 10 μL), 50 mM citrate-phosphate buffer pH 5.0 (10 μL) and the enzyme solution (30 μL). After incubation of the reaction mixture at 35 °C for 10 min with shaking at 850 rpm, 0.1 M Na_2CO_3 (1 mL) was added to the reaction mixture. The released *p*-nitrophenol was determined spectrophotometrically at 420 nm. One unit of enzymatic activity was defined as the amount of enzyme releasing 1 μmol of *p*-nitrophenol per minute in 50 mM citrate-phosphate buffer at pH 5.0.

Enzymatic Preparation of Colorimetric Substrate *p*-Nitrophenyl Rutinoside

The colorimetric substrate *p*-nitrophenyl rutinoside (not commercially available) was prepared by the glycosylation of *p*-nitrophenyl glucopyranoside using α-L-rhamnosidase (*A. terreus*) and a free rhamnose yielding the title substrate, however with very low yields (ca 3%) [17]. Therefore, we developed another method for the preparation of this compound, taking advantage of the high transglycosylation activity of rutinosidase from *A. niger* towards phenolic substances [17,24]. The following protocol was used: *p*-nitrophenol (100 mg) and rutin (200 mg) were dissolved in DMSO (1.1 mL) and crude rutinosidase (cultivation medium of transformed *P. pastoris*; 6.4 mL, 0.4 U/mL) was added. The pH was adjusted to 3.0 and the mixture was incubated at 40 °C for 24 h. The reaction was monitored by TLC and HPLC. The reaction was stopped by boiling for 10 min, filtered, the pH was adjusted to 7.5–7.7 and unreacted *p*-nitrophenol was removed by extraction (3 × 30 mL EtOAc). The aqueous phase was evaporated, the residue was dissolved in 20% *v*/*v* methanol in water (2 mL) and loaded into a Sephadex LH-20 column (eluted with 20% *v*/*v* methanol). Fractions containing the product (TLC, Silicagel 60 F_{254}, Merck; *n*-propanol/H_2O/NH_4OH = 7:2:1, *v*/*v*/*v*) were pooled and evaporated to yield 47 mg (32% related to rutin) of the title compound.

4.8. Bioconversion of Rutin to Quercetin

4.8.1. Analytical Scale

Purified recombinant rutinosidase from *P. pastoris* was used for the optimization of biotransformation conditions.

pH Optimum

Rutin (3, 200 g/L) was suspended in the enzyme solution, which contained 0.2 M glycine-buffer and the pure enzyme (0.2 U/mL). The buffer pH was adjusted to 2.0; 2.5; 3.0; 3.5; 4.0; 4.5 or 5.0. The solution (3 mL) was incubated at 35 °C with shaking (750 rpm) for 7 h. Samples were taken every 30 min for the HPLC analysis.

Enzyme Activity

Rutin (3, 200 mg/mL) was suspended in the solution (3 mL) containing 0.2 M glycine-buffer pH 3.5 and the pure enzyme of various activities (0.021; 0.2; 0.4 or 0.8 U/mL). The reaction mixture was incubated at 35 °C with shaking (750 rpm) for 24 h. The reaction was monitored by HPLC. 1 U is defined as the amount of enzyme converting 1 μM of substrate within 1 min under the given conditions.

Temperature Optimum

Rutin (3, 200 g/L) was suspended in a solution (10 mL) containing 0.2 M glycine-buffer pH 3.5 and pure enzyme (0.2 U/mL). The reaction mixture was incubated at 10 °C; 20 °C; 35 °C; 40 °C or 50 °C with shaking (750 rpm) for 24 h. Samples were taken every 30 min for the HPLC analysis.

Thermostability

The crude medium (*P. pastoris*) containing rutinosidase was adjusted to pH 3.0 and was incubated at 50 °C for 150 min. Every 10 min, samples (30 μL) were taken and the enzyme activity was measured using the above procedure.

4.8.2. Scale-Up of Bioconversion of Rutin to Quercetin

Bioconversion Scale-Up Using Crude Wild-Type Rutinosidase (*A. niger*)

For the scale-up process, the centrifuged crude medium from the cultivation of the production strain *A. niger* containing rutinosidase (0.3 U/mL) was used. The crude medium was diluted twice with water and the pH was adjusted to 3.5. The medium itself had a good buffering capacity. Rutin (3; 50 g/L; 100 g/L; 150 g/L or 200 g/L) was suspended in the buffered enzyme solution, incubated at 40 °C with shaking (750 rpm) and the reaction conversion was monitored by HPLC.

Bioconversion Scale-Up Using Crude Recombinant Rutinosidase (*P. pastoris*)

For the scale-up process, the crude medium (prepared in a fermenter—see scale up) from transformed *P. pastoris* (containing typically 0.4 U/mL) was adjusted to 0.2 U/mL with water and the pH was adjusted to 3.5 with H_3PO_4. The medium itself had a good buffering capacity. Rutin (3; 200 g/L; 300 g/L or 450 g/L) was suspended in the buffered enzyme solution, incubated at 40 °C with shaking (750 rpm) and the reaction conversion was monitored by HPLC. Rutin was completely consumed after 24 h; then the reaction was stopped by heating to 99 °C for 5 min. The reaction mixture was then centrifuged (5000× *g*, 10 min at 20 °C) or filtered.

4.8.3. Microscopic Monitoring of the Solid State Biocatalysis

The progress of bioconversion of rutin to quercetin catalyzed by the crude recombinant rutinosidase (rutin 100 g/L) was monitored with an OLYMPUS CX41 optical microscope equipped with an OLYMPUS U-CMAD3 digital camera (Olympus Europa SE & Co., KG, Hamburg, Germany).

4.9. Analytical methods

Analytical HPLC

HPLC analyses were performed in a Shimadzu Prominence LC analytical system (Shimadzu, Kyoto, Japan) consisting of the following Shimadzu components: LC-20AD binary HPLC pump, CTO-10AS column oven, SIL-20ACHT cooling autosampler, CBM-20A system controller and SPD-20MA diode array detector. The sample (20 μL) was dissolved in DMSO (150 μL) and analyzed by a Chromolith Performance RP-18e column (100 × 3 mm, Merck, Germany) coupled with an RP-18e guard cartridge kit (5 × 4.6 mm, Merck, Darmstadt, Germany). Binary gradient elution was used: mobile phase A (*v*/*v*): 5% acetonitrile, 0.1% formic acid; mobile phase B (*v*/*v*): 80% acetonitrile, 0.1% formic acid; gradient: 7–25% B for 0–3 min, 30% B for 3–5 min; 7% B for 5–7.5 min. The flow rate was 1.5 mL/min at 25 °C and the injection volume was 0.1 μL; peaks were detected at 360 nm. Retention times [min]: rutin, 2.380; quercetin, 3.964. The authenticity of the compounds, for example, rutin and quercetin, was re-confirmed by MS spectra using LC-MS with commercial samples. Structure of rutinose was confirmed by ^{1}H and ^{13}C NMR [17].

Glycerol and methanol concentrations were measured by HPLC with an Agilent Technologies 1220 Infinity LC apparatus combined with an Agilent Technologies 1260 Infinity RI detector (Agilent Technologies, Waldbronn, Germany), a WATREX Polymer IEX H form 8 μm, 250 × 8 mm as the main column and a WATREX Polymer IEX H form 8 μm, 40 × 8 mm (WATREX, Prague, Czech Republic) as a guard column, at a flow rate of 0.8 mL/min of 9 mM sulfuric acid at 45 °C.

5. Conclusions

We present a novel biotechnological process for the production of the bioactive flavonoid quercetin from the inexpensive starting material rutin. The main feature of the process can be described as "solid-state" enzymatic conversion with both substrate and product being mostly in an undissolved state. The enzymatic reaction itself thus proceeds in the microenvironment solution of the reactants. This new concept of "solid-state biocatalysis" takes advantage of the low water solubility of both the starting material and product. Using this approach with a specific diglycosidase, rutinosidase, quercetin is obtained from the quantitative bioconversion of rutin. Recombinant rutinosidase is used as a crude enzyme directly from a bioreactor cultivation (after biomass centrifugation). In addition, the procedure leads to the formation of the valuable and commercially unavailable disaccharide rutinose as a second product of the reaction. This work demonstrates an elegant approach to biotransformation under mild conditions in the absence of co-solvents and toxic chemicals, potentially transferable to conversions of other natural compounds and food additives. Thanks to the new biocatalytic methodology void of any harmful chemicals, the new process, which results in the products quercetin and rutinose, can be declared as environmentally friendly or a "bio" quality product on the market.

Supplementary Materials: Supplementary materials can be found at http://www.mdpi.com/1422-0067/20/5/1112/s1.

Author Contributions: Conceptualization, V.K. and J.K.; investigation, L.P.; J.K.; M.K.; K.M.; data curation, J.K.; L.P.; writing—original draft preparation, V.K.; writing—review and editing, P.B.; M.R.; M.K.; supervision, V.K., funding acquisition, V.K.; M.R.

Funding: This research was funded by Czech Science Foundation grant number 18-00150S and the EDRF project from Slovak Research Agency grant number ITMS 26230120009.

Acknowledgments: We wish to thank Ben Watson-Jones MEng for providing linguistic assistance.

Conflicts of Interest: The authors declare no conflict of interest.

Abbreviations

GRAS	Generally-Recognized-As-Safe
DMSO	Dimethylsulfoxide
DCW	Dry Cell Weight
DO	Dissolved Oxygen
BMMY	Buffered Methanol-Complex Medium
BMGY	Buffered Glycerol-Complex Medium
BMGH	Buffered Minimal Glycerol Medium
BMMH	Buffered Minimal Methanol Medium
BSM	Basal Salt Medium
PTM_1	Trace Salt Solution
YNB	Yeast Nitrogen Base
vvm	volume of air per volume of liquid per minute

References

1. Russo, M.; Spagnuolo, C.; Tedesco, I.; Bilotto, S.; Russo, G.L. The flavonoid quercetin in disease prevention and therapy: Facts and fancies. *Biochem. Pharmacol.* **2012**, *83*, 6–15. [CrossRef] [PubMed]
2. Almeida, A.F.; Borge, G.I.A.; Piskula, M.; Tudose, A.; Tudoreanu, L.; Valentová, K.; Williamson, G.; Santos, C.N. Bioavailability of quercetin in humans with a focus on inter-individual variation. *Compr. Rev. Food Sci. Food Saf.* **2018**, *17*, 714–731. [CrossRef]
3. Valentová, K.; Vrba, J.; Bancířová, M.; Ulrichová, J.; Křen, V. Isoquercitrin: Pharmacology, toxicology, and metabolism. *Food Chem. Toxicol.* **2014**, *68*, 267–282. [CrossRef] [PubMed]
4. Gullon, B.; Lu-Chau, T.A.; Moreira, M.T.; Lema, J.M.; Eibes, G. Rutin: A review on extraction, identification and purification methods, biological activities and approaches to enhance its bioavailability. *Trends Food Sci. Technol.* **2017**, *67*, 220–235. [CrossRef]
5. D'Andrea, G. Quercetin: A flavonol with multifaceted therapeutic applications? *Fitoterapia* **2015**, *106*, 256–271. [CrossRef] [PubMed]
6. Özyurt, H.; Çevik, Ö.; Özgen, Z.; Ozden, A.S.; Cadirci, S.; Elmas, M.A.; Ercan, F.; Gören, M.Z.; Şener, G. Quercetin protects radiation-induced DNA damage and apoptosis in kidney and bladder tissues of rats. *Free Rad. Res.* **2014**, *48*, 1247–1255. [CrossRef] [PubMed]
7. Schimmer, O.; Kruger, A.; Paulini, H.; Haefele, F. An evaluation of 55 commercial plant-extracts in the Ames mutagenicity test. *Pharmazie* **1994**, *49*, 448–451. [PubMed]
8. Andres, S.; Pevny, S.; Ziegenhagen, R.; Bakhiya, N.; Schäfer, B.; Hirsch-Ernst, K.I.; Lampen, A. Safety aspects of the use of quercetin as a dietary supplement. *Mol. Nutr. Food Res.* **2017**, *62*, 1700447. [CrossRef] [PubMed]
9. Carullo, G.; Cappello, A.R.; Frattaruolo, L.; Badolato, M.; Armentano, B.; Aiello, F. Quercetin and derivatives: Useful tools in inflammation and pain management. *Future Med. Chem.* **2017**, *9*, 79–93. [CrossRef] [PubMed]
10. Biler, M.; Biedermann, D.; Valentová, K.; Křen, V.; Kubala, M. Quercetin and its analogues: Optical and acido-basic properties. *Phys. Chem. Chem. Phys.* **2017**, *19*, 26870–26879. [CrossRef] [PubMed]
11. Levdanskii, V.A.; Polezhaeva, N.I.; Kuznetsov, B.N. Method for Producing Quercetin, Involves Oxidizing Larch Wood with Sodium Nitrite and Hydrolysis by Superheated Vapor and Subjecting to Rapid Decompression. Russian Patent RU2182906-C1, 13 June 2001.
12. Zhao, Z. Production of Quercetin by Mixing Inorganic Strong Acid and Water, Adding Rutin, Boiling, Filtering, Hydrolyzing, Washing Cake with Water, Drying, Mixing Cake with Methanol, Filtering, Washing Cake with Methanol and Water, and Drying. Chinese Patent CN104387357-A, 19 December 2011.
13. Wang, J.; Zhao, L.-L.; Sun, G.-X.; Liang, Y.; Wu, F.-A.; Chen, Z.; Cui, S. A comparison of acidic and enzymatic hydrolysis of rutin. *Afr. J. Biotechnol.* **2011**, *10*, 1460–1466. [CrossRef]
14. Cui, X.; Wang, Z. New Rutin Hydrolase Used for Preparing Quercetin From Rutin, Prepared by Extracting Tartary Buckwheat Powder Using Acetate Buffer, and Processing Extract in Anion Exchange Chromatography and Gel Chromatography. Chinese Patent CN101787361-A, 11 January 2010.
15. Nam, H.K.; Hong, S.-H.; Shin, K.-C.; Oh, D.-K. Quercetin production from rutin by a thermostable beta-rutinosidase from *Pyrococcus furiosus*. *Biotechnol. Lett.* **2012**, *34*, 483–489. [CrossRef] [PubMed]

16. Rebroš, M.; Pilniková, A.; Šimčíková, D.; Weignerová, L.; Stloukal, R.; Křen, V.; Rosenberg, M. Recombinant α-L-rhamnosidase of *Aspergillus terreus* immobilization in polyvinylalcohol hydrogel and its application in rutin derhamnosylation. *Biocatal. Biotransform.* **2013**, *31*, 329–334. [CrossRef]
17. Šimčíková, M.; Kotik, M.; Weignerová, L.; Halada, P.; Pelantová, H.; Adamcová, K.; Křen, V. α-L-Rhamnosyl-β-D-glucosidase (rutinosidase) from *Aspergillus niger*: Characterization and synthetic potential of a novel diglycosidase. *Adv. Synth. Catal.* **2015**, *357*, 107–117. [CrossRef]
18. Salamin, K.; Sriranganadane, D.; Léchenne, B.; Jousson, O.; Monod, M. Secretion of an endogenous subtilisin by *Pichia pastoris* strains GS115 and KM71. *Appl. Environ. Microbiol.* **2010**, *76*, 4269–4276. [CrossRef] [PubMed]
19. Markošová, K.; Weignerová, L.; Rosenberg, M.; Křen, V.; Rebroš, M. Upscale of recombinant α-L-rhamnosidase production by *Pichia pastoris* MutS strain. *Front. Microbiol.* **2015**, *6*, 1140. [CrossRef] [PubMed]
20. Bojarová, P.; Kulik, N.; Slámová, K.; Hubálek, M.; Kotik, M.; Cvačka, J.; Pelantová, H.; Křen, V. Selective β-*N*-acetylhexosaminidase from *Aspergillus versicolor*—A tool for producing bioactive carbohydrates. *Appl. Microbiol. Biotechnol.* **2019**, *103*, 1737–1753. [CrossRef] [PubMed]
21. Weignerová, L.; Marhol, P.; Gerstorferová, D.; Křen, V. Preparatory production of quercetin-3-β-D-glucopyranoside using alkali-tolerant thermostable α-L-rhamnosidase from *Aspergillus terreus*. *Bioresour. Technol.* **2012**, *115*, 222–227. [CrossRef] [PubMed]
22. Faury, G.; Molinari, J.; Rusova, E.; Mariko, B.; Raveaud, S.; Huber, P.; Velebný, V.; Robert, A.M.; Robert, L. Receptors and aging: Structural selectivity of the rhamnose-receptor on fibroblasts as shown by Ca^{2+}-mobilization and gene-expression profiles. *Arch. Gerontol. Geriatr.* **2011**, *53*, 106–112. [CrossRef] [PubMed]
23. Vishniac, W.; Santer, M. The thiobacilli. *Bacteriol. Rev.* **1957**, *21*, 195–213. [PubMed]
24. Bassanini, I.; Krejzová, J.; Panzeri, W.; Křen, V.; Monti, D.; Riva, S. A sustainable one-pot two enzyme synthesis of naturally occurring arylalkyl glucosides. *ChemSusChem* **2017**, *10*, 2040–2045. [CrossRef] [PubMed]

International Journal of
Molecular Sciences

MDPI

Article

Cell Wall Anchoring of a Bacterial Chitosanase in *Lactobacillus plantarum* Using a Food-Grade Expression System and Two Versions of an LP × TG Anchor

Mai-Lan Pham [1,†], Anh-Minh Tran [1,2,†], Geir Mathiesen [3], Hoang-Minh Nguyen [4,*] and Thu-Ha Nguyen [1,*]

1 Food Biotechnology Laboratory, Department of Food Science and Technology, BOKU-University of Natural Resources and Life Sciences, Vienna, Muthgasse 18, 1190 Vienna, Austria; mailanpham.22@gmail.com (M.-L.P.); anh.tran@boku.ac.at (A.-M.T.)
2 Department of Biology, Faculty of Fundamental Sciences, Ho Chi Minh City University of Medicine and Pharmacy, 217 Hong Bang, Ho Chi Minh City, Vietnam
3 Faculty of Chemistry, Biotechnology and Food Science, Norwegian University of Life Sciences (NMBU), P.O. Box 5003, N-1432 Ås, Norway; geir.mathiesen@nmbu.no
4 Department of Biotechnology, The University of Danang-University of Science and Technology, 54 Nguyen Luong Bang, Da Nang 550000, Vietnam
* Correspondence: nhminh@dut.udn.vn (H.-M.N.); thu-ha.nguyen@boku.ac.at (T.-H.N.)
† These authors contributed equally to this work.

Received: 9 May 2020; Accepted: 23 May 2020; Published: 27 May 2020

Abstract: Lactic acid bacteria (LAB) have attracted increasing interest recently as cell factories for the production of proteins as well as a carrier of proteins that are of interest for food and therapeutic applications. In this present study, we exploit a lactobacillal food-grade expression system derived from the pSIP expression vectors using the *alr* (alanine racemase) gene as the selection marker for the expression and cell-surface display of a chitosanase in *Lactobacillus plantarum* using two truncated forms of a LP × TG anchor. CsnA, a chitosanase from *Bacillus subtilis* 168 (ATCC23857), was fused to two different truncated forms (short-S and long-L anchors) of an LP × TG anchor derived from Lp_1229, a key-protein for mannose-specific adhesion in *L. plantarum* WCFS1. The expression and cell-surface display efficiency driven by the food-grade *alr*-based system were compared with those obtained from the *erm*-based pSIP system in terms of enzyme activities and their localisation on *L. plantarum* cells. The localization of the protein on the bacterial cell surface was confirmed by flow cytometry and immunofluorescence microscopy. The highest enzymatic activity of CsnA-displaying cells was obtained from the strain carrying the *alr*-based expression plasmid with short cell wall anchor S. However, the attachment of chitosanase on *L. plantarum* cells via the long anchor L was shown to be more stable compared with the short anchor after several repeated reaction cycles. CsnA displayed on *L. plantarum* cells is catalytically active and can convert chitosan into chito oligosaccharides, of which chitobiose and chitotriose are the main products.

Keywords: surface display; cell wall anchor; *Lactobacillus plantarum*; whole-cell biocatalyst

1. Introduction

Chitin, which can be found in the outer exoskeleton of arthropods, such as crabs, lobster, shrimp, and in the fungal cell wall, is the second most abundant biopolymer after lignocellulose in nature [1]. An enormous amount of chitin is annually disposed into the sea, which requires a need for bioremediation or recycling of the chitin biomass [2]. Chitosan is a natural nontoxic biopolymer, which can be obtained from the partially de-acetylation of chitin under alkaline conditions and is composed of β-1,4-*N*-acetyl-D-glucosamine (GlcNac) and D-glucosamine (GlcN) [3]. Despite its biological activities, its applications were limited due to its high molecular weight and solubility at acidic pH [4]. However, the depolymerisation products from chitosan, so-called chito-oligosaccharides (CHOS), which possess potential properties for pharmaceutical and medical applications, are of interest due to their nontoxic and high solubility properties [3,4]. CHOS can be produced by enzymatic hydrolysis using chitosanases or chitosan *N*-acetyl-glucosaminohydrolases (EC3.2.1.132) [3,5]. These enzymes catalyse the endohydrolysis β-1,4-glycosidic bond between D-glucosamine of chitosan and have been found in various organisms including Gram-positive bacteria, Gram-negative bacteria, and fungi [5–9]. Among them, the chitosanases from family GH46, especially *Bacillus subtilis* 168, showed the potential for bioconversion of chitosan to produce CHOS [5,10]. The chitosanase from *B. subtilis* 168, which is encoded by *csnA* gene, was demonstrated to be efficiently produced and secreted in *Lactobacillus plantarum* using the inducible promoter-based pSIP expression vectors containing either an antibiotic resistance or the *alr* gene as selection markers [11].

Recently, bacterial cell-surface display has become an attractive strategy for the development of whole-cell biocatalysts. By fusing with an anchoring motif, the protein of interest can be simultaneously synthesized and subsequently displayed on the bacterial cells. Obtained cells harbouring fusion target protein/enzyme after fermentation can be directly used as biocatalysts for the reaction processes. This concept offers known advantages such as the immobilisation of enzymes with a significant reduction of production cost and utilisation of the bacterial biomass as the immobilisation matrix instead of traditional carrier materials. With a long history of use in food industry and possession of GRAS (generally regarded as safe) status, *L. plantarum* has been exploited as host for cell-surface display of heterologous proteins [12–21]. In our previous study, we reported cell-surface display of a chitosanase via a lipoprotein anchor (Lp_1261) and an LP × TG motif-containing a cell wall anchor from a cell surface adherence protein (Lp_2578) of *L. plantarum* WCFS1 [22], using the original *erm*-based pSIP expression system [23,24]. It was shown that surface display of the chitosanase in lactobacilli using the Lp_2578 derived cell wall anchor resulted in higher enzyme activity than *Lactobacillus* strains with Lp_1261 derived lipoprotein anchor [22].

Nevertheless, the use of antibiotic resistance gene as the selection marker in original pSIP vectors [23,25], which requires the supplement of erythromycin in cultivation media, might lead to the limitation of this system in some applications, e.g., production of food ingredients and additives. Therefore, a food-grade selection marker based on complementation would be a more useful alternative [26]. We have previously developed a food-grade, complementation-based host/marker expression system derived from the pSIP expression vectors using the *alr* (alanine racemase) gene as the selection marker for intracellular expression of β-galactosidases in *L. plantarum* [27]. In the present study, we exploit this food-grade expression system for the expression and cell-surface display of a chitosanase derived from *B. subtilis* 168 (ATCC23857) in *L. plantarum*. We have constructed two variants of LP × TG anchor derived from Lp_1229 with different lengths of the linker between the anchoring motif and the chitosanase to investigate the effect of anchor length on the enzymatic activity of chitosanase displaying cells. The expression and cell-surface display efficiency driven by the food-grade system were compared with those obtained from the *erm*-based pSIP system in term of enzyme activities and their localisation on *L. plantarum* cells.

2. Results

2.1. Expression of Chitosanase (CsnA) in L. plantarum

Four new expression vectors were constructed to display a chitosanase from *Bacillus subtilis* 168 on the surface of *L. plantarum* (Figure 1A–C) using both *erm*-based and *alr*-based expression systems (erythromycin and alanine racemase as selection markers, respectively) in lactobacilli. The enzyme was C-terminally anchored to the cell wall using two different truncated forms, namely S (Short) and L (long) anchors, of an LP × TG anchor derived from the Lp_1229 sequence, which encoding a protein containing predicted mucus binding domains in *L. plantarum* [28,29]. For the immunodetection of the recombinant proteins, a Myc-tag was fused C-terminally to the CsnA sequence. The short anchor S (85 residues) consisting of the LP × TG motif, which is the actual consensus sequence LPQTNE in *L. plantarum* [30,31] (in bold, underlined, Figure 1A), and a linker of 47 residues upstream of the LP × TG motif were fused C-terminally to CsnA-Myc sequence using a valine residue (V in bold, Figure 1A). The longer anchor L (194 residues) contains a linker of 156 residues upstream of the LP × TG motif, which consists of one Mub B2-like domain of 69 residues (mucus binding domain; in italics, Figure 1A) [28,32] that is separated from consensus sequence LPQTNE by 77 residues, and 10 residues upstream of Mub B2-like domain (Figure 1A). The CsnA-Myc sequence was fused to L anchor at threonine residue (T in bold, Figure 1A). For efficient secretion, a signal peptide (SP) derived from Lp_0373, which was previously shown to efficiently secrete heterologous proteins in *L. plantarum* [33–35], was fused to CsnA through the *Sal*I linker (Figure 1A).

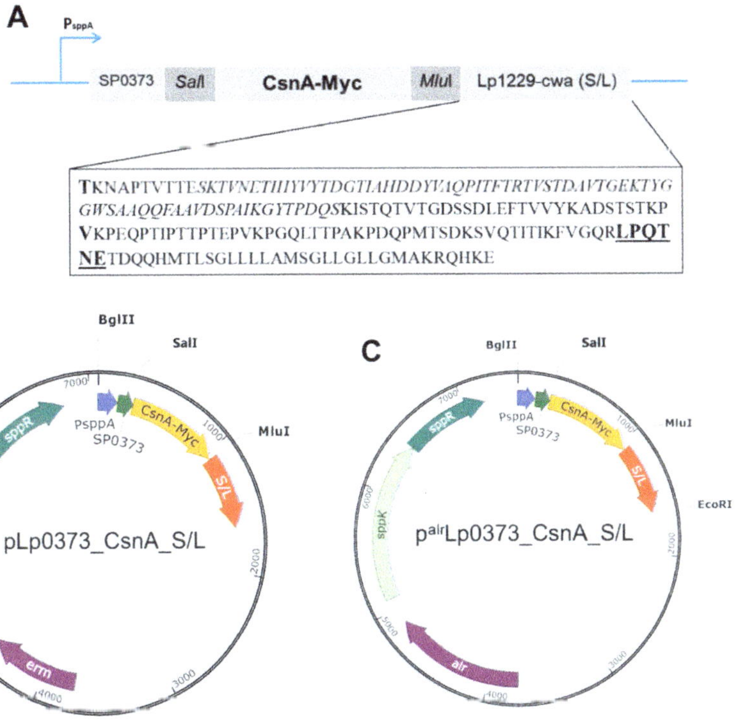

Figure 1. The expression cassette (**A**), *erm*-based and *alr*-based expression vectors (**B**,**C**) of C-terminal cell-wall anchor of chitosanase (CsnA) in *L. plantarum*. (**A**) Myc tag sequence was fused to the C-terminus of CsnA for protein detection. All parts of the expression cassette can be easily exchangeable using the indicated restriction enzymes (*Bgl*II, *Sal*I, *Mul*I, *EcoR*I). The signal peptide (SP) derived from Lp_0373 was fused to CsnA via *Sal*I site. The short anchor S (85 residues) consists of a linker of 47 residues between the LP × TG motif, which is the consensus sequence LPQTNE in *L. plantarum* [30,31] (bold, underlined), and the C-terminus of CsnA-Myc with a valine as the fusing point (V in bold). The long anchor L (194 residues) contains a linker of 156 residues upstream of the LP × TG motif including one Mub B2-like domain of 69 residues (mucus binding domain; in italics) fused to CsnA-Myc with a threonine residue (T in bold). (**B**) Schematic overview of the plasmid encoding CsnA fused to a LP × TG cell wall anchor using different lengths of Lp_1229 derived anchor and Lp_0373 derived SP

for secretion. The plasmid contains an erythromycin antibiotic resistance gene (*erm*) as selection marker. (**C**) Schematic overview of the plasmid encoding CsnA fused to a LP × TG cell wall anchor using different lengths of Lp_1229 derived anchor and Lp_0373 derived SP for secretion. The plasmid contains alanine racemase gene (*alr*) as selection marker. See text for more details.

All engineered plasmids, pLp0373_CsnA_S/L and p^{alr}Lp0373_CsnA_S/L, were then transformed into *L. plantarum* WCFS1 or D-alanine auxotroph strain, *L. plantarum* TLG02 [27], respectively. To analyse the production of the chitosanase, Western blot analysis of the crude, cell-free extracts was performed using anti-Myc antibodies for detection of the enzyme. Figure 2 shows that all four recombinant strains harbouring the plasmids of both expression systems produced the expected proteins. The *L. plantarum* WCFS1 strain harbouring an empty vector (pEV) (lane 1, Figure 2A) and the *L. plantarum* TLG02 strain without plasmid (lane 1, Figure 2B) used as the negative controls did not produce any target protein as expected, while the strain harbouring pSIP_CsnAcwa2 (lane 2, Figure 2A), in which the chitosanase from *B. sutbtilis* 168 is fused to an LP × TG anchor derived from the Lp_2578 (cwa2) [12,22], was used as a positive control. Figure 2A,B show the presence of chitosanases with two cell wall anchors S/L at expected molecular weights of ~39 kDa and ~52 kDa, respectively (Figure 2A, lanes 3–4 for the strains harbouring pLp0373_CsnA_S/L; Figure 2B, lanes 2–3 for strains harbouring p^{alr}Lp0373_CsnA_S/L, respectively).

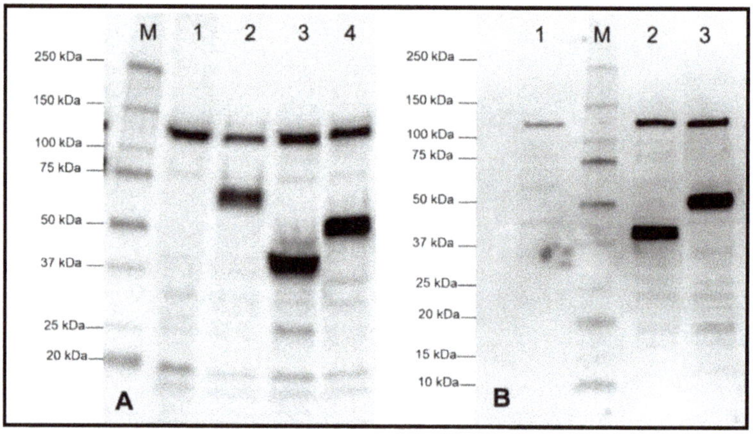

Figure 2. Western blot analysis of cell-free extracts from transformed and induced *L. plantarum* cells harbouring various expression plasmids. (**A**) *erm*-based expression system (1) pEV, empty vector as negative control; (2) pSIP_CsnAcwa2: inducible expression vector for cell wall anchoring of CsnA using cwa2 derived from the Lp_2578 protein as positive control [22]; (3) pLp0373_CsnA_S (expected protein size 39 kDa); (4) pLp0373_CsnA_L (expected protein size 52 kDa). (**B**) *alr*-based expression system (1) *L. plantarum* TLG02 as negative control [27]; (2) p^{alr}Lp0373_CsnA_S (expected protein size 39 kDa); (3) p^{alr}Lp0373_CsnA_L (expected protein size 52 kDa). M indicates molecular mass markers (Biorad).

2.2. *Enzymatic Activity of CsnA-Displaying Cells*

Chitosanase activities of living recombinant lactobacillal cells were measured using 0.5% (w/v) of chitosan as the substrate to evaluate the functionality of the surface-displayed enzymes. Figure 3A,B shows the time course of the cultivations of CsnA-displaying *L. plantarum* strains harboring *erm*-based and *alr*-based expression plasmids, respectively, and the highest levels of chitosanase-displaying activities (U/g dry cell weight) of recombinant bacteria from both systems were obtained at 4 h after induction (OD_{600} ~ 1.0–1.2). A significant decrease in activities (U/g dry cell weight; U/g DCW) was observed as the cultivation was extended after 8 h of induction. The highest enzymatic activities of CsnA-displaying cells of 1975 U/g DCW and 2160 U/g DCW were obtained from the strains carrying the plasmids with short cell wall anchor S, pLp0373_CsnA_S and p^{alr}Lp0373_CsnA_S, respectively (Figure 3A,B). On the other hand, the highest enzymatic activities of CsnA-displaying cells obtained from the strains carrying the plasmids with long cell wall anchor L, pLp0373_CsnA_L and p^{alr}Lp0373_CsnA_L, were 1300 U/g DCW and 1690 U/g DCW, respectively (Figure 3A,B). It was also confirmed that the enzymatic activities obtained from CsnA-displaying cells were indeed from surface-anchored chitosanase as no enzymatic activities were detected from *Lactobacillus* cells carrying the empty plasmid pEV (negative control) or from the host strains.

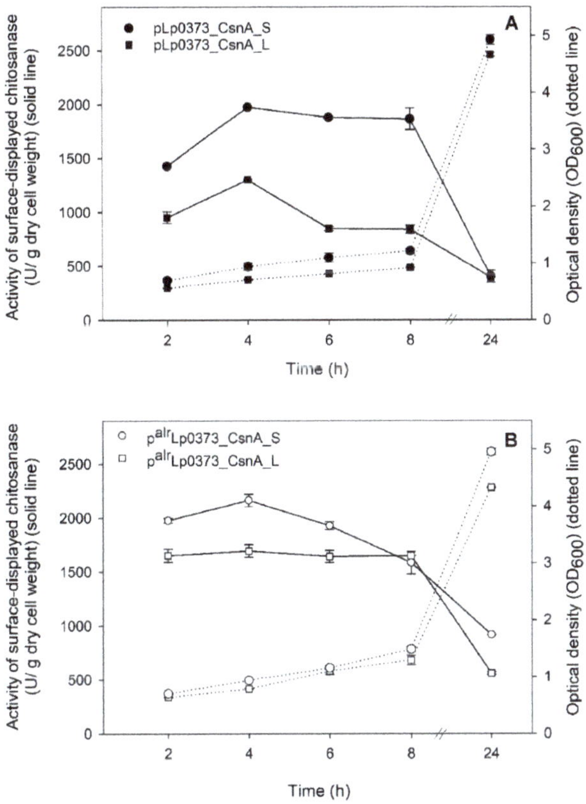

Figure 3. Enzymatic activity of CsnA displaying cells. Time course of cultivations of CsnA-displaying *L. plantarum* recombinant strains harboring the plasmids of *erm*-based expression system (**A**) and *alr*-based expression system (**B**) in MRS medium at 37 °C. Experiments were performed in triplicates, and the standard deviation was always less than 5%.

2.3. Surface Localization of CsnA in L. plantarum and the Stability of CsnA-Displaying Cells

Even though significant enzymatic activities of CsnA-displaying cells were obtained from the strains carrying the plasmids with short cell wall anchor S, only slight shift in the fluorescence signal was detected for the strain harbouring the plasmid of *erm*-based expression system pLp0373_CsnA_S and we could not observe a clear shift in the fluorescence signal for the strain harbouring the plasmid of *alr*-based expression system p^{alr}Lp0373_CsnA_S compared to the control strain (Figure 4A). On the other hand, flow cytometry analysis confirmed surface display of the chitosanase for the strains carrying the plasmids with long cell wall anchor L, pLp0373_CsnA_L and p^{alr}Lp0373_CsnA_L, as indicated by the clear shifts in the fluorescence signals observed for these strains compared to the negative control strains (Figure 4A). Interestingly, immunofluorescence microscopy clearly confirmed the surface localization of the Myc-tag in all recombinant strains carrying the plasmids with both short and long cell wall anchors. However the number of CsnA-displaying cells from the strains carrying the plasmids with short cell wall anchor S, pLp0373_CsnA_S and p^{alr}Lp0373_CsnA_S, were notably lower compared to the strains carrying the plasmids with long cell wall anchor L (Figure 4B), even though the same number of cells in all samples were used at the beginning of the experiments.

These results are supported by the observations of the stability of CsnA-displaying cells. We measured the enzyme activities of surface-displayed chitosanase of *Lactobacillus* cells during four repeated cycles with a washing step between the cycles to remove proteins released from lysed cells. Interestingly, CsnA-displaying cells from the strains carrying the plasmids with short anchor S showed less stable after repeated cycles compared to the strains carrying the plasmids with long anchor L. The enzymatic activities of *Lactobacillus* cells harbouring the plasmids with S anchor, pLp0373_CsnA_S and p^{alr}Lp0373_CsnA_S, decreased significantly as indicated by activity losses of ~50% and 35%, respectively, after four assay/washing cycles (Figure 5A,B). The chitosanase displaying cells via L anchor of *erm*- and *alr*-based expression systems retained more than 70% and ~90% after four assay/washing cycles, respectively, of their initial chitosanase activities, indicating that these enzyme-displaying cells can be reused for several rounds of biocatalysis at 37 °C (Figure 5A,B). Hence, the observation that a lower number of cells harbouring pLp0373_CsnA_S and p^{alr}Lp0373_CsnA_S exposed the fluorescence signals in immunofluorescence microscopy analysis can be explained by the fact that a high number of cells lost the anchored chitosanase during the preparation of the cells prior to the analysis due to the unstable attachment of the enzyme to the cell wall.

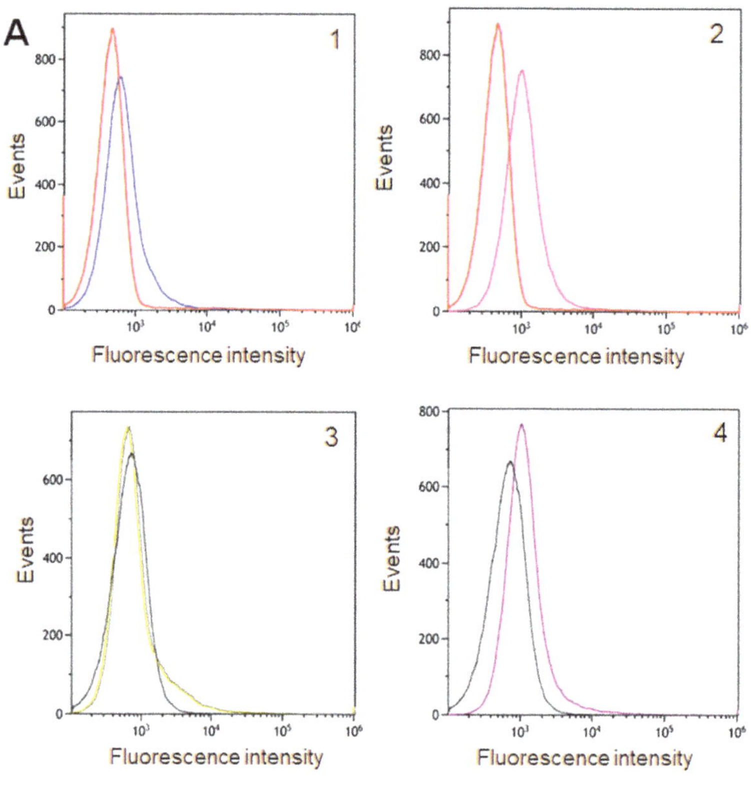

Figure 4. *Cont.*

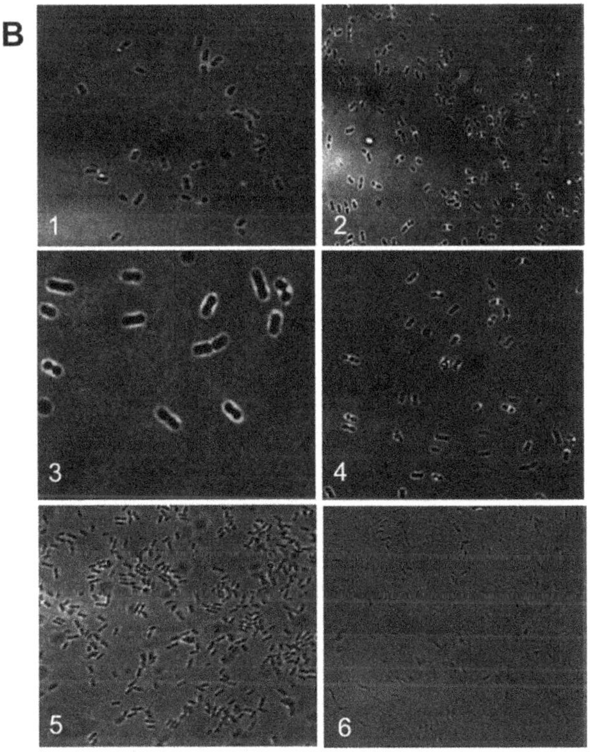

Figure 4. Surface localisation of chitosanase in *L. plantarum* cells analysed by flow cytometry (**A**) and immunofluorescence microscopy (**B**). The *L. plantarum* strains are denoted by different numbers in the flow cytometry histograms (**A**) and in the micrographs (**B**): pLp0373_CsnA_S (**A1**,**B1**); pLp0373_CsnA_L (**A2**,**B2**); p^{alr}Lp0373_CsnA_S (**A3**,**B3**); p^{alr}Lp0373_CsnA_L (**A4**,**B4**); *L. plantarum* WCFS1 harboring an empty vector pEV (red line in **A1**–**A2**; **B5**) and *L. plantarum* TLG02 (black line in **A3**–**A4**; **B6**) were used as negative controls.

Furthermore, the thermal stability of *L. plantarum* TLG02 cells carrying the plasmid p^{alr}Lp_0373_CsnA_L at various temperatures was investigated. It was shown that chitosanase displaying cells are very stable at the storage temperature of −20 °C with a half-life of ~9 months (data not shown). The activity of CsnA-anchored cells was also stable at 37 °C. In fact, the displaying cells lost 50% of initial chitosanase activity only after five weeks. CsnA-displaying cells retained 50% of initial enzyme activity for ~24 h at 50 °C (data not shown), which is of great interest from an application point of view for chitosan conversion.

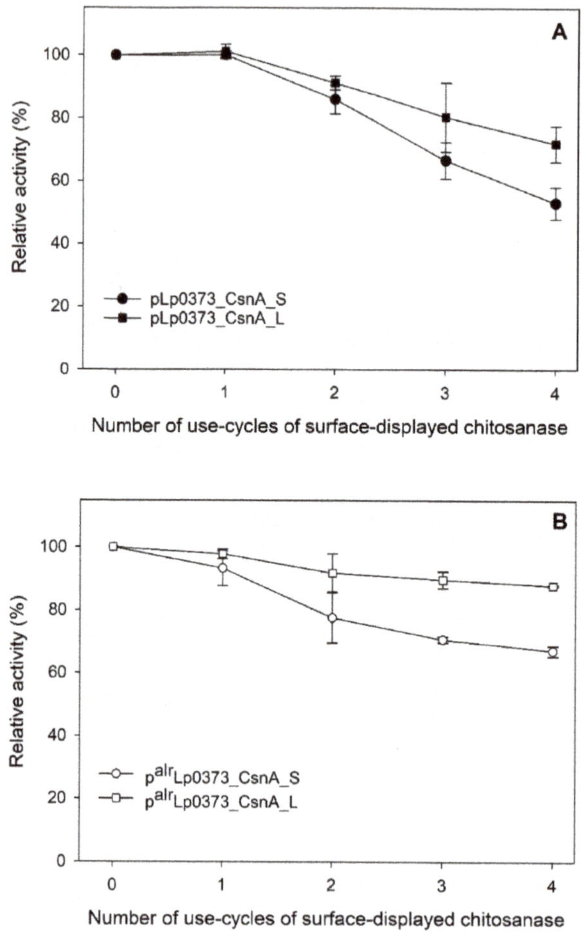

Figure 5. Enzymatic activity of CsnA displaying cells of repeated activity measurements. 0 indicates the first measurement after harvesting, while 1, 2, 3, 4 indicate the number of replications. (**A**) *L. plantarum* cells harbouring the *erm*-based expression plasmids (**B**) *L. plantarum* cells harbouring the *alr*-based expression plasmids. Experiments were performed in triplicates, and the standard deviation was always less than 5%.

2.4. Chitosan Conversion and Products Analysis by Thin Layer Chromatography (TLC)

Due to their high stability, *L. plantarum* TLG02 cells harbouring the plasmid p^{alr}Lp_0373_CsnA_L were selected for enzymatic hydrolysis of chitosan at 37 °C and 50 °C. Chitosan solutions were prepared from two different sources, low molecular weight and chitosan from crab shells medium molecular weight, as described in the materials and methods. The conversion of chitosan to chito-oligosaccharides (CHOS) by CsnA-displaying cells was analyzed by TLC analysis. Analysis of product formation by TLC revealed that surface-displayed chitosanase can degrade low and medium molecular weight chitosan into CHOS (Figure 6A–D). However, high molecular weight chitosan was hardly hydrolysed (data not shown). Chitobiose (C2) and chitotriose (C3) are main products from the conversions of both substrates (Figure 6D). Chitotetraose (C4) was found in the reaction mixtures from the conversions of both substrates after 5 min of the conversions at 50 °C (Figure 6C,D) and it was further degraded with prolonged incubation time in the reaction with low molecular weight substrate (Figure 6C).

Interestingly, chitopentaose (C5) was present in the reaction mixture with medium molecular weight substrate at the beginning of the conversion at 50 °C and was then hydrolysed after 15 min (Figure 6D).

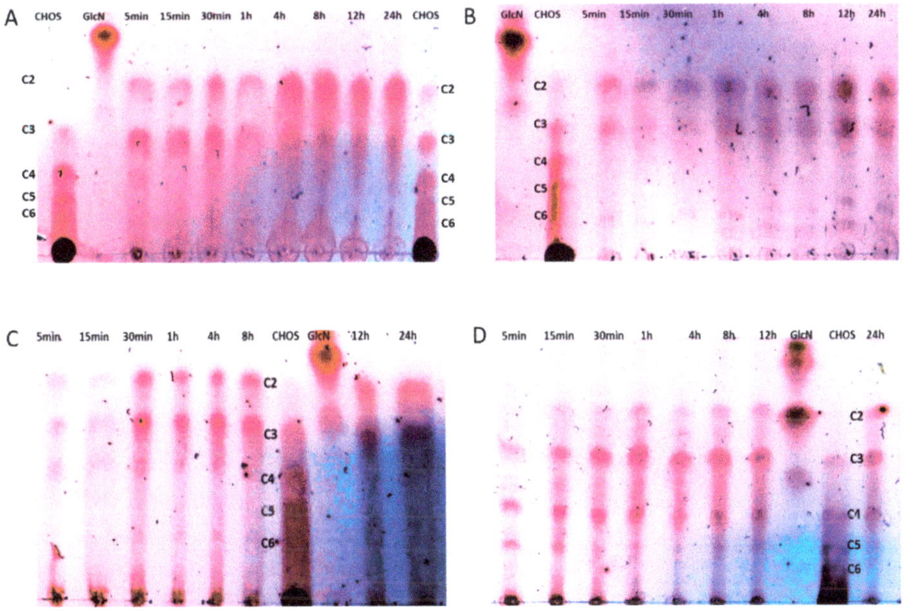

Figure 6. Formation of chito-oligosaccharides (CHOS) from chitosan (0.5% w/v) by chitosanase-displaying *L. plantarum* cells harboring p^{alr}Lp0373_CsnA_L at 37 °C (**A**,**B**) and 50 °C (**C**,**D**) using two different substrates: chitosan low molecular weight (**A**,**C**) and chitosan from crab shells medium molecular weight (**B**,**D**) as analysed by thin layer chromatography (TLC). Standards: commercial chito-oligosaccharides (TCI Deutschland GmBH, Eschborn, Germany) which contains chitobiose (**C2**), chitotriose (**C3**), chitotetraose (**C4**), chitopentose (**C5**), chitohexaose (**C6**) and D-glucosamine (GlcN).

3. Discussion

In present study, we constructed an expression system for the surface display of a chitosanase from *B. subtilis* in *L. plantarum* based on the pSIP603 expression vector with *alr* gene as the selection marker [27] and the newly constructed system was compared with the original *erm*-based expression system. Two truncated forms S/L with corresponding lengths of 85 and 195 residues from an LP × TG anchor derived from the Lp_1229 sequence were employed to anchor the chitosanase to the cell wall of *L. plantarum*. The S-anchor contains the predicted proline-rich region (PxxP repetitive sequences) located upstream of the C-terminal LPQTNE anchoring motif, whereas the L-anchor extends further and consists of the Mub B2 like domain (mucus binding domain) (Figure 1), which is supposed to generate flexibility of the protein chain and play an important role in the adherence of LAB to the mucus layer covering the epithelial cells of the intestine [29,36]. It was expected that, with the inclusion of the Mub B2-like domain in the linker of L-anchor, CsnA would be more exposed at the surface. Surprisingly, recombinant strains harbouring both the *erm*- and the *alr*-based expression plasmids with the short anchor S resulted in significantly higher chitosanase activities than the long anchor L ($p < 0.001$ with both *alr*- and *erm* based systems). The activity of surface-displayed enzyme decreases as the length of the anchor is increased. Nevertheless, the attachment of chitosanase on *L. plantarum* cells via the short anchor S was shown to be significantly less stable compared with the long anchor L after several repeated reaction cycles ($p < 0.05$ with *erm*-based system and $p < 0.001$ with *alr*-based system). It could be an explanation for the observations from flow cytometry and immunofluorescence microscopy analyses that the fluorescence signals obtained from the cells harbouring the plasmids with the long anchor L were stronger than those with the short anchor S. Several washing steps required during the preparation for immuno-detection may result in the release of chitosanase from lactobacillal cells carrying the plasmids with the short anchor S. As a result, the long anchor L demonstrates more stable surface localization of chitosanase among these two truncated versions of the LP × TG anchor derived from the Lp_1229 sequence. Furthermore, the food-grade *alr*-based pSIP expression system yields significantly higher production of displayed CsnA than the conventional *erm*-based system ($p < 0.01$ with S anchor and $p < 0.001$ with L anchor).

Interestingly, the attachment of chitosanase on *L. plantarum* cells using these Lp_1229 derived anchors, especially the short anchor S, was more efficient than the cell wall anchor (cwa2) derived from the Lp_2578 protein ($p < 0.001$), of which the highest chitosanase activity was 1360 U per gram dry cell weight (~1.7 mg per g dry cell weight) [22]. The long anchor L has almost the same length as cwa2 (194 residues) and the strain carrying the *erm*-based plasmid, pLp0373_CsnA_L, showed no significant difference in displayed-chitosanase activity, whereas the strain carrying the *alr*-based plasmid, p^{alr}Lp0373_CsnA_L, had significantly higher displayed chitosanase activity compared to the cwa2 ($p < 0.01$). However, it should be noted that the plasmid constructed for surface display of CsnA using cwa2 contain N-terminal signal peptide derived from the gene encoding Lp_3050 [22], which is different from the signal peptide Lp0373 used in this study.

4. Materials and Methods

4.1. Bacterial Strains and Chemicals

The bacterial strains used in this study are listed in Table 1. *Lactobacillus plantarum* WCFS1, isolated from human saliva as described by Kleerebezem et al. [31] was originally obtained from NIZO Food Research (Ede, The Netherlands) and maintained in the culture collection of the Norwegian University of Life Sciences (NMBU), Ås, Norway. *L. plantarum* WCFS1 and TLG02 were grown in deMan, Rogosa and Shape (MRS) broth without/with the addition of 200 µg/mL of D-alanine (Sigma), respectively, at 37 °C without agitation. *Escherichia coli* HST08 (Clontech, Mountain View, CA, USA) and *E. coli* MB2159 used in the transformation experiments involving the subcloning of DNA fragments were cultivated in Luria-Bertani (LB) broth without/with the addition of 200 µg/mL of D-alanine, respectively, at 37 °C with agitation. The agar plates were prepared by adding 1.5% agar to the respective media. When needed, erythromycin was supplemented to media to final concentrations of 5 µg/mL for *L. plantarum* and 200 µg/mL for *E. coli*.

Table 1. Strains and plasmids used in this study.

Strain or Plasmid	Relevant Characteristic (s)	Reference Source
Strains		
L. plantarum		
WCFS1	wild type, host strain	[31]
TLG02	Δ*alr*, D-alanine auxotroph, food-grade expression host	[27]
E. coli		
HST08	cloning host	Clontech
MB2159	D-alanine auxotroph, cloning host	[37]
Plasmids		
pLp0373_ManB_S	Erm^r; pSIP401 derivate encoding the Lp_0373 signal peptide translationally fused to *manB-myc*, followed by the short cell wall anchor (S) from Lp_1229	(unpublished)
pLp0373_ManB_L	Erm^r; pSIP401 derivate encoding the Lp_0373 signal peptide translationally fused to *manB-myc*, followed by the long cell wall anchor (L) from Lp_1229	(unpublished)

Table 1. *Cont.*

Strain or Plasmid	Relevant Characteristic (s)	Reference Source
pSIP409-CsnA-native	Erm^r; *spp*- based expression vector pSIP409 for expression of *csnA* with native signal peptide	[11]
pEV	Erm^r; pSIP401 derivative, empty vector, no signal sequence, no *csnA* (negative control)	[13]
pSIP603-GusA	Erm^r; pSIP401 derivative, *gusA* controlled by P_{sppA}, *alr* replaced *erm*	[27]
pLp0373_CsnA_S	Erm^r; pLp0373_ManB_S derivative with *csnA-myc* instead of *manB-myc*	This study
pLp0373_CsnA_L	Erm^r; pLp0373_ManB_L derivative with *csnA-myc* instead of *manB-myc*	This study
p^{alr}Lp0373_CsnA_S	pSIP603 derivative with SPLp0373-*csnA-myc*-S instead of *gusA*	This study
p^{alr}Lp0373_CsnA_L	pSIP603 derivative with SPLp0373-*csnA-myc*-L instead of *gusA*	This study

4.2. DNA Manipulation

Plasmids were isolated from *E. coli* strains using the Monarch plasmid miniprep kit (New England Biolabs, Frankfurt am Main, Germany), following the instructions of the manufacturers. PCR products and digested fragments were purified using the Monarch DNA Gel extraction kit (New England Biolabs, Frankfurt am Main, Germany) and the DNA amounts were estimated by Nanodrop 2000 (Thermo Fisher Scientific, Waltham, MA, USA). DNA amplifications were performed using Q5® High-Fidelity DNA Polymerase (New England Biolabs, Frankfurt am Main, Germany) and the primers listed in Table 2. Sequences of PCR generated fragments were verified by DNA sequencing performed by a commercial provider (Microsynth, Vienna, Austria). The digestion by restriction enzymes (New England Biolabs) and the ligation of DNA fragments by In-fusion HD Cloning kit (Clontech, Mountain View, CA, USA) were performed following the instructions of the manufacturers. The plasmids with erythromycin resistance gene (*erm*) and the plasmids with alanine racemase gene (*alr*) as the selection markers were transformed into *E. coli* HST08 and *E. coli* MB2159 chemical competent cells, respectively, following the manufacturers' protocols for obtaining the plasmids in sufficient amounts. The constructed plasmids were transformed into electrocompetent cells of *L. plantarum* WCFS1 or *L. plantarum* TLG02 WCFS1 according to the protocol of Aukrust and Blom [38].

Table 2. Primers used in this study.

Primer	Sequence [a] 5′ → 3′	Restriction Site Underlined
Fwd1_CsnA_SalI	TGCTTCATCAGTCGAC*GCGGGACTGAATAAAGATC*	*Sal*I
Fwd2_CsnA-BglII	ATTACAGCTCCAGATCT*ACCGGTGGGCC*	*Bgl*II
Rev1_CsnA	TGAGATGAGTTTTTGTTC*GTCGACACATCCTTTGATTAC*	
Rev2_CsnA	CAGATCCTCTTC*TGAGATGAGTTTTTGTTCGTCGACAGA*	
Rev3_CsnA_MluI_S	CTGGTTTAACACGCGT*CAGATCCTCTTCTGAGATG*	*Mlu*I
Rev4_CsnA_MluI_L	GAGCATTCTTGGTACGCGT*CAGATCCTCTTC*	*Mlu*I
Rev5_CsnA_S/L_EcoRI	GGGGTACCGAATTC*AAGCTTCTACTCTTTGTGCTGTC*	*Eco*RI

[a] The nucleotides in italics are the positions that anneal to the DNA of the target gene.

4.3. Plasmid Construction

Two truncated forms S and L of the cell wall anchor were derived from Lp_1229, a 1010 amino acid protein encoded by *msa* gene involved in mannose specific adhesion in *L. plantarum* cell wall, which contains LPQTNE motif. The short anchor S comprises 85 C-terminal residues, whereas the long anchor L has the total length of 195 amino residues containing one Mub B2-like domain (mucin binding domain). The anchoring sequences (S, L) used in this study were taken from pLp0373_ManB_S and pLp0373_ManB_L (Table 1), which are derivatives of the pSIP401 vector that has been developed for inducible gene expression in lactobacilli [24]. These pLp0373_ManB_S and pLp0373_ManB_L were used for the construction of the expression plasmids in this study and they all contain a bacterial mannanase gene *manB*, which was fused N-terminally to the signal peptide Lp_0373 [33] and C-terminally to a 30-bp fragment encoding the *myc* tag (GAACAAAAACTCATCTCAGAAGAGGATCTG), followed by the anchoring sequences S and L, respectively.

For construction of the erythromycin-based expression plasmids pLp0373_CsnA_S and pLp_0373_CsnA_L, the *csnA-myc* fragment was generated by three PCR steps using pSIP409_CsnA_native [11] as the template: PCR1 with primers Fwd1_CsnA_SalI and Rev1_CsnA to introduce a N-terminal *Sal*I sites, PCR2 with primers Fwd1_CsnA_SalI and Rev2_CsnA, PCR3 with primers Fwd1_CsnA_SalI and Rev3_CsnA_MluI_S or Rev4_CsnA_MluI_L, which are compatible with short anchor S or long anchor L encoding fragments, respectively, to introduce a C-terminal *Mlu*I sites. The resulting PCR fragments were ligated into the *Sal*I/*Mlu*I digested plasmids: pLp0373_ManB_S and pLp0373_ManB_L using In-Fusion HD Cloning kit (Clontech, Mountain View, CA, USA) resulting in two plasmids pLp0373_CsnA_S and pLp_0373_CsnA_L, respectively (Figure 1A,B).

For the construction of the food-grade expression plasmids, the expression cassettes, P_{sppA}-SPLp0373-*csnA-myc*-S/L (~1.3–1.6 kb, Figure 1A) containing the promoter P_{sppA} (of pSIP401), the signal peptide Lp_0373, *csnA* gene fused to the *myc* tag followed by one of the anchoring sequences S/L were amplified using the primer pair Fwd2_CsnA_BglII and Rev5_CsnA_S/L_EcoRI and the newly constructed plasmids pLp0373_CsnA_S and pLp0373_CsnA_L as templates. The resulting PCR fragments were then ligated into ~5.8 kb *Bgl*II-*Eco*RI digested fragment of pSIP603 vector, in which the erythromycin resistance gene (*erm*) was replaced by the alanine racemase gene (*alr*) as the selection marker [27], using In-Fusion HD Cloning kit (Clontech, Mountain View, CA, USA) yielding two food-grade expression plasmids named p^{alr}Lp0373_CsnA_S and p^{alr}Lp_0373_CsnA_L (Figure 1C).

4.4. Gene Expression in *L. plantarum*

The expression plasmids pLp0373_CsnA_S and pLp0373_CsnA_L were constructed in *E. coli* HST08 before electroporation into *L. plantarum* WCFS1 competent cells and transformants were selected on MRS agar plates containing 5 µg/mL erythromycin. To generate the food-grade expression strains, the food-grade expression plasmids p^{alr}Lp0373_CsnA_S and p^{alr}Lp0373_CsnA_L were constructed in *E. coli* MB2159 before electroporation into electro-competent *L. plantarum* TLG02, a D-alanine auxotroph expression host and the selection of transformants was performed on MRS agar plates.

Gene expression was carried out by diluting the overnight cultures of *L. plantarum* strains harbouring the plasmids in 100 mL of fresh pre-warm MRS broth (for *erm*-based systems, 5 µg/mL of erythromycin was added) to an OD_{600} of ~0.1, and incubated at 37 °C without agitation. The cells were induced at an OD_{600} of 0.3 by adding the peptide pheromone IP-673 [39] to final concentration of 25 ng/mL. Cells were harvested at OD_{600} of ~1.0–1.2 by centrifugation (4000× *g*, 4 °C, 10 min), washed twice with phosphate buffered saline (PBS) containing 137 mM NaCl, 2.7 mM KCl, 2 mM KH_2PO_4, and 10 mM Na_2HPO_4 (pH 7.4), and then re-suspended in PBS buffer.

4.5. Enzymatic Activity Measurement

Chitosanase activity was determined as described previously [10,22] with some modifications. Chitosanase-displaying cells were collected from the cultures by centrifugation at 4000× *g* for 10 min at 4 °C. Cell pellets obtained from cultures were washed twice with PBS and re-suspended in 200 µL of PBS. Chitosan (PT Biotech Surindo, Jawa Barat, Indonesia) was completely dissolved in 1% (w/v) of acetic acid at 80 °C for 30 min before adjusting the pH of the solution to 5.5 with 1 M NaOH. The reaction was conducted with 100 µL of an enzyme-displaying cells suspension in PBS buffer and 400 µL of a 0.5% (w/v) chitosan solution at 37 °C for 5 min with mixing at 800 rpm. The cells and the supernatant were separated by centrifugation (4000× *g*, 4 °C, 2 min). The amount of reducing sugar released in the supernatant of the enzymatic reaction was determined by the dinitroasalicylic acid (DNS) assay. DNS assay was carried out with 100 μL of the reaction supernatant mixed with 100 μL of DNS solution at 99 °C for 10 min, followed by cooling on ice for 5 min. The mixture was then diluted with 800 μL of de-ionised water before measuring the absorbance at 540 nm using 1–5 μmol/mL of D-glucosamine as standards. One unit of chitosanase activity was defined as the amount of enzyme releasing 1 μmol of reducing sugars (or reducing end equivalents) per minute under the given conditions.

4.6. Western Blotting

The cells obtained from 50 mL of cultures were disrupted with glass-beads (170 μm; Sigma Aldrich, Darmstadt, Germany) using the Precelly 24 glass bead mill (PEQLAB Biotechnology GmbH, Erlangen, Germany) and cell-free extracts (crude extracts) were obtained after 5 min of centrifugation at 10,000× *g* and 4 °C. Protein concentrations in the cell-free extracts were measured by Bradford assay [40] and separated on SDS-PAGE gels before being transferred to a nitrocellulose membrane using the Trans-Blot Turbo Transfer system (Bio-Rad Laboratories, Hercules, CA, USA) following the instructions of the manufacturer. After nonspecific protein interactions were blocked by incubating the membrane with 50 mL of 1% BSA dissolved in Tris-buffered saline-Tween 20 (TBS-T) for one hour on the shaker at room temperature, the membrane was immediately incubated with 1 μL of monoclonal murine anti-Myc antibody (Invitrogen, Carlsbad, CA, USA; diluted 1:5000) in TBS-T buffer containing 0.5% of BSA at 4 °C overnight. After three times washing with 15 mL of TBS-T buffer, the membrane was incubated with 2.5 μL of a secondary antibody, which was polyclonal rabbit anti-mouse antibody conjugated with horseradish peroxidase (HRP) (Dako, Glostrup, Denmark), diluted 1:2000 in TBS-T buffer containing 0.5% BSA for 1 h in the dark at room temperature. Before visualization, the membrane was rinsed again three times with 15 mL of TBS-T, following by incubation with a Clarity Western ECL Blotting Substrate (Bio-Rad Laboratories, Hercules, CA, USA). The protein bands were visualized by the ChemiDoc™ XRS+ imaging system (Bio-Rad Laboratories, Hercules, CA, USA).

4.7. Flow Cytometry Analysis

Cell staining for flow cytometry was carried out as previously described [22] with some modifications. One mL of cell culture (OD_{600} of ~0.5) was obtained 2 h after induction, and the cells were incubated in 50 μL PBS containing 2% of BSA (PBS-B) with 0.1 μL monoclonal anti-Myc antibody (Invitrogen, Carlsbad, CA, USA; diluted 1:250 in PBS-B) at room temperature for 40 min. Subsequently, the cell suspension was centrifuged at 5000× *g* for 3 min at 4 °C, washed three times with 500 μL PBS-B, and then incubated with 0.2 μL anti-mouse IgG antibody (Alexa Fluor 488 conjugated; Cell Signalling Technology, Frankfurt am Main, Germany) diluted 1:500 in 50 μL PBS-B for 40 min at room temperature, in the dark. After collecting the cells by centrifugation (4000× *g*, 3 min at 4 °C) and washing three times with 500 μL PBS-B, the stained cells were resuspended in 100 μL of PBS and analysed using a Gallios Flow cytometer (Beckman Coulter, Brea, CA, USA), following the manufacturer's instructions. The data were analysed by Kaluza Analysis software (Beckman Coulter, Brea, CA, USA)

4.8. Indirect Immunofluorescence Microscopy Analysis

Cell fixation and staining for indirect immunofluorescence microscopy were carried out as previously described [22,41–43] with some modifications. One mL of cell culture (OD_{600} of ~0.5) was harvested 2 h after induction by centrifugation at 4000× *g* for 10 min at 4 °C, and re-suspended in 100 μL PBS (pH 7.4). The cells were fixed with ethanol to final concentration of 70% (w/w) for 1 h at −20 °C [41,42]. Fixed cells were washed twice with PBS and was resuspended in 100 μL of PBS. An amount of 30 μL of fixed cells was transferred onto a microscope slide and the cells were absorbed on the slide for 2 h at room temperature until the trace was dried. Fixed cells were then incubated with 100 μL of PBS containing 10% of BSA for 1 h to block nonspecific protein interactions. Subsequently, the cells were stained with anti-Myc antibody (Invitrogen, Carlsbad, CA, USA) and anti-mouse IgG antibody (Alexa Fluor 488 conjugated; Cell Signalling Technology, Frankfurt am Main, Germany) as described above. The stained cells were washed three times with 100 μL of PBS (5 min for each washing step) and the slide was mounted with 5 μL of mounting medium containing 50% glycerol in PBS [43]. The stained cells were observed under a Leica DMI6000B 'Live cell' wide-field fluorescence microscope (Leica Microsystems; Wetzlar, Germany) using the 488-nm argon laser line. The fluorescence detection window was set between 505 nm and 550 nm and the images were acquired with a PL APO 63×/1.40 oil immersion objective.

4.9. Catalytic Stability and Thermal Stability of Chitosanase Displaying Cells

Chitosanase displaying cells were collected from cultures by centrifugation at 4000× *g* for 10 min at 4 °C. Cell pellets were washed twice with PBS and re-suspended in 100 μL of PBS and chitosanase activities were measured at 37 °C as described above. This procedure was repeated for several cycles of activity measurements with intermediate two washing steps to determine the number of use cycles of surface displayed chitosanase.

In order to determine the thermal stability of chitosanase displaying cells at various temperatures (−20, 37, and 50 °C), *L. plantarum* TLG02 cells harbouring p^{alr}Lp_0373_CsnA_L were collected from 20 mL culture by centrifugation at 4000× *g* for 10 min at 4 °C and re-suspended in 1 mL of PBS prior to incubation at these temperatures. At certain time intervals, the enzymatic activity of chitosanase-displaying cells was measured using chitosan as the substrate under standard assay conditions. The half-life value ($\tau_{\frac{1}{2}}$) of activity was determined when residual activity reaches 50%.

4.10. Chitosan Conversion

Conversion of 0.5% (w/v) chitosan low molecular weight and chitosan from crab shells medium molecular weight, which were prepared in 1% (w/v) of acetic acid at 80 °C for 30 min before adjusting the pH of the solution to 5.5 with 1 M NaOH, catalysed by surface displayed chitosanase was carried out on a 2-mL scale with 5 U of chitosanase per mL of reaction mixture at 37 °C and 50 °C for 24 h. Surface-displayed chitosanase was obtained from the expression strain *L. plantarum* TLG02 harbouring the plasmid p^{alr}Lp_0373_CsnA_L. Agitation was applied at 150 rpm and the samples were taken at regular intervals. The reactions were stopped by heating the samples at 99 °C for 5 min prior to further analyses.

4.11. Thin Layer Chromatography (TLC) Analysis

TLC was performed by high performance TLC (HPTLC) silicagel plate (Kiselgel 60 F245, Merck, Kenilworth, NJ, USA) as previously described [10,44] with some modifications. Approximately 0.5 μL of samples were applied to the plate and eluted twice in ascending mode with an iso-propanol/25% ammonia/water mixture (7:1:2). Thymol reagent was used for visualization. A commercial chito-oligosaccharides (TCI Deutschland GmBH, Eschborn, Germany) with a mixture of C2–C6, which contains chitobiose, chitotriose, chitotetraose, chitopentose, chitohexaose, and D-glucosamine (GlcN) (Sigma Aldrich, Darmstadt, Germany) were used as standards.

4.12. Statistical Analysis

All measurements were conducted in triplicates, and the standard deviation was always less than 5%. Student's *t* test was used for the comparison of data.

5. Conclusions

We have demonstrated the successful anchoring of a bacterial chitosanase onto the cell surface of *L. plantarum* using a food-grade lactobacillal expression system and two truncated forms of the LP × TG cell wall anchor derived from the Lp_1229 sequence. CsnA displayed on *L. plantarum* cells is catalytically active and can convert chitosan into chito-oligosaccharides, of which chitobiose and chitotriose are the main products. The successful development of the food-grade *alr*-based expression system, which overcomes certain drawbacks of the original *erm*-based system, for the surface display of an active chitosanase in *L. plantarum* will certainly result in a stable, food-grade, whole-cell biocatalyst that could be of interest for the production of oligosaccharides of prebiotic potential.

Author Contributions: Conceptualization, G.M., H.-M.N. and T.-H.N.; Data curation, M.-L.P., G.M. and T.-H.N.; Formal analysis, M.-L.P. and A.-M.T.; Investigation, M.-L.P. and A.-M.T.; Methodology, M.-L.P., H.-M.N. and T.-H.N.; Supervision, T.-H.N.; Writing—original draft, M.-L.P.; Writing—review & editing, G.M. and T.-H.N. All authors have read and agreed to the published version of the manuscript.

Funding: This research is funded by the Vietnam National Foundation for Science and Technology Development (NAFOSTED) under grant number 106.02-2019.339 for H.-M.N. and by the Austrian Science Fund (FWF Project V457-B22) for T.-H.N. Open Access of this publication is funded by the Austrian Science Fund (FWF).

Acknowledgments: The authors thank Karin Kohlweiss (Institute for Biotechnology, BOKU) and Monika Debreczeny (Immaging Center, VIBT, BOKU) for their assistance and guidance to M.-L.P. in FACS and Immunofluorescence microscopy analyses. M.-L.P. thanks the European Commission for the Erasmus Mundus scholarship under the ALFABET project. A.-M.T. is thankful for the Ernst Mach—ASEA UNINET scholarship granted by the OeAD—Austrian Agency for International Cooperation in Education and Research and financed by the Austrian Federal Ministry of Science, Research and Economy. Research documented in this publication was supported by the ASEAN-European Academic University Network (ASEA-UNINET).

Conflicts of Interest: The authors declare no conflict of interest.

References

1. Khoushab, F.; Yamabhai, M. Chitin research revisited. *Mar. Drugs* **2010**, *8*, 1988–2012. [CrossRef] [PubMed]
2. Gooday, G.W. The ecology of chitin degradation. In *Advances in Microbial Ecology*; Marshall, K.C., Ed.; Springer: Boston, MA, USA, 1990; Volume 11, pp. 387–430. [CrossRef]
3. Aam, B.B.; Heggset, E.B.; Norberg, A.L.; Sørlie, M.; Vårum, K.M.; Eijsink, V.G.H. Production of chitooligosaccharides and their potential applications in medicine. *Mar. Drugs* **2010**, *8*, 1482–1517. [CrossRef] [PubMed]
4. Hoell, I.A.; Vaaje-Kolstad, G.; Eijsink, V.G.H. Structure and function of enzymes acting on chitin and chitosan. *Biotechnol. Genet. Eng. Rev.* **2010**, *27*, 331–366. [CrossRef] [PubMed]
5. Pechsrichuang, P.; Lorentzen, S.B.; Aam, B.B.; Tuveng, T.R.; Hamre, A.G.; Eijsink, V.G.H.; Yamabhai, M. Bioconversion of chitosan into chito-oligosaccharides (CHOS) using family 46 chitosanase from *Bacillus subtilis* (BsCsn46A). *Carbohydr. Polym.* **2018**, *186*, 420–428. [CrossRef] [PubMed]
6. Ando, A.; Saito, A.; Arai, S.; Usuda, S.; Furuno, M.; Kaneko, N.; Shida, O.; Nagata, Y. Molecular characterization of a novel family-46 chitosanase from *Pseudomonas* sp. A-01. *Biosci. Biotechnol. Biochem.* **2008**, *72*, 2074–2081. [CrossRef] [PubMed]
7. Wang, S.L.; Peng, J.H.; Liang, T.W.; Liu, K.C. Purification and characterization of a chitosanase from *Serratia marcescens* TKU011. *Carbohyd. Res.* **2008**, *343*, 1316–1323. [CrossRef]
8. Zhang, X.Y.; Dai, A.L.; Zhang, X.K.; Kuroiwa, K.; Kodaira, R.; Shimosaka, M.; Okazaki, M. Purification and characterization of chitosanase and exo-β-D-glucosaminidase from a koji mold, *Aspergillus oryzae* IAM2660. *Biosci. Biotechnol. Biochem.* **2000**, *64*, 1896–1902. [CrossRef]
9. Zhou, J.; Liu, X.; Yuan, F.; Deng, B.; Yu, X. Biocatalysis of heterogenously-expressed chitosanase for the preparation of desirable chitosan oligosaccharides applied against phytopathogenic fungi. *ACS Sustain. Chem. Eng.* **2020**, *8*, 4781–4791. [CrossRef]
10. Pechsrichuang, P.; Yoohat, K.; Yamabhai, M. Production of recombinant *Bacillus subtilis* chitosanase, suitable for biosynthesis of chitosan-oligosaccharides. *Bioresour. Technol.* **2013**, *127*, 407–414. [CrossRef]
11. Sak-Ubol, S.; Namvijitr, P.; Pechsrichuang, P.; Haltrich, D.; Nguyen, T.-H.; Mathiesen, G.; Eijsink, V.G.; Yamabhai, M. Secretory production of a beta-mannanase and a chitosanase using a *Lactobacillus plantarum* expression system. *Microb. Cell Fact.* **2016**, *15*, 81. [CrossRef]
12. Fredriksen, L.; Mathiesen, G.; Sioud, M.; Eijsink, V.G. Cell wall anchoring of the 37-kilodalton oncofetal antigen by *Lactobacillus plantarum* for mucosal cancer vaccine delivery. *Appl. Environ. Microbiol.* **2010**, *76*, 7359–7362. [CrossRef] [PubMed]
13. Fredriksen, L.; Kleiveland, C.R.; Hult, L.T.; Lea, T.; Nygaard, C.S.; Eijsink, V.G.; Mathiesen, G. Surface display of N-terminally anchored invasin by *Lactobacillus plantarum* activates NF-kappaB in monocytes. *Appl. Environ. Microbiol.* **2012**, *78*, 5864–5871. [CrossRef] [PubMed]

14. Moraïs, S.; Shterzer, N.; Lamed, R.; Bayer, E.A.; Mizrah, I. A combined cell-consortium approach for lignocellulose degradation by specialized *Lactobacillus plantarum* cells. *Biotechnol. Biofuels* **2014**, *7*, 112. [CrossRef] [PubMed]
15. Kuczkowska, K.; Mathiesen, G.; Eijsink, V.G.; Oynebraten, I. *Lactobacillus plantarum* displaying CCL3 chemokine in fusion with HIV-1 Gag derived antigen causes increased recruitment of T cells. *Microb. Cell Fact.* **2015**, *14*, 169. [CrossRef] [PubMed]
16. Kuczkowska, K.; Kleiveland, C.R.; Minic, R.; Moen, L.F.; Øverland, L.; Tjåland, R.; Carlsen, H.; Lea, T.; Mathiesen, G.; Eijsink, V.G.H. Immunogenic properties of *Lactobacillus plantarum* producing surface-displayed *Mycobacterium tuberculosis* antigens. *Appl. Environ. Microbiol.* **2017**, *83*, e02782-16. [CrossRef] [PubMed]
17. Kuczkowska, K.; Copland, A.; Øverland, L.; Mathiesen, G.; Tran, A.C.; Paul, M.J.; Eijsink, V.G.H.; Reljic, R. Inactivated *Lactobacillus plantarum* carrying a surface-displayed Ag85B-ESAT-6 fusion antigen as a booster vaccine against *Mycobacterium tuberculosis* infection. *Front. Immunol.* **2019**, *10*, 1588. [CrossRef]
18. Kuczkowska, K.; Øverland, L.; Rocha, S.D.C.; Eijsink, V.G.H.; Mathiesen, G. Comparison of eight *Lactobacillus* species for delivery of surface-displayed mycobacterial antigen. *Vaccine* **2019**, *37*, 6371–6379. [CrossRef]
19. Bober, J.R.; Nair, N.U. Galactose to tagatose isomerization at moderate temperatures with high conversion and productivity. *Nat. Commun.* **2019**, *10*, 4548. [CrossRef]
20. Pham, M.L.; Tran, A.M.; Kittibunchakul, S.; Nguyen, T.T.; Mathiesen, G.; Nguyen, T.-H. Immobilization of β-galactosidases on the *Lactobacillus* cell surface using the peptidoglycan-binding motif LysM. *Catalysts* **2019**, *9*, 443. [CrossRef]
21. Nguyen, H.M.; Pham, M.L.; Stelzer, E.M.; Plattner, E.; Grabherr, R.; Mathiesen, G.; Peterbauer, C.K.; Haltrich, D.; Nguyen, T.-H. Constitutive expression and cell-surface display of a bacterial β-mannanase in *Lactobacillus plantarum*. *Microb. Cell Fact.* **2019**, *18*, 76. [CrossRef]
22. Nguyen, H.M.; Mathiesen, G.; Stelzer, E.M.; Pham, M.L.; Kuczkowska, K.; Mackenzie, A.; Agger, J.W.; Eijsink, V.G.H.; Yamabhai, M.; Peterbauer, C.K.; et al. Display of a beta-mannanase and a chitosanase on the cell surface of *Lactobacillus plantarum* towards the development of whole-cell biocatalysts. *Microb. Cell Fact.* **2016**, *15*, 169. [CrossRef] [PubMed]
23. Sorvig, E.; Grönqvist, S.; Naterstad, K.; Mathiesen, G.; Eijsink, V.G.H.; Axelsson, L. Construction of vectors for inducible gene expression in *Lactobacillus sakei* and *L. plantarum*. *FEMS Microbiol. Lett.* **2003**, *229*, 119–126. [CrossRef]
24. Sorvig, E.; Mathiesen, G.; Naterstad, K.; Eijsink, V.G.H.; Axelsson, L. High-level, inducible gene expression in *Lactobacillus sakei* and *Lactobacillus plantarum* using versatile expression vectors. *Microbiology* **2005**, *151*, 2439–2449. [CrossRef] [PubMed]
25. Remiger, A.; Eijsink, V.G.H.; Ehrmann, M.A.; Sletten, K.; Nes, I.F.; Vogel, R.F. Purification and partial amino acid sequence of plantaricin 1.25α and 1.25β, two bacteriocins produced by *Lactobacillus plantarum* TMW1.25. *J. Appl. Microbiol.* **1999**, *86*, 1053–1058. [CrossRef] [PubMed]
26. Peterbauer, C.; Maischberger, T.; Haltrich, D. Food-grade gene expression in lactic acid bacteria. *Biotechnol. J.* **2011**, *6*, 1147–1161. [CrossRef] [PubMed]
27. Nguyen, T.T.; Mathiesen, G.; Fredriksen, L.; Kittl, R.; Nguyen, T.-H.; Eijsink, V.G.; Haltrich, D.; Peterbauer, C.K. A food-grade system for inducible gene expression in *Lactobacillus plantarum* using an alanine racemase-encoding selection marker. *J. Agric. Food Chem.* **2011**, *59*, 5617–5624. [CrossRef]
28. Pretzer, G.; Snel, J.; Molenaar, D.; Wiersma, A.; Bron, P.A.; Lambert, J.; de Vos, W.M.; van der Meer, R.; Smits, M.A.; Kleerebezem, M. Biodiversity-based identification and functional characterization of the mannose-specific adhesin of *Lactobacillus plantarum*. *J. Bacteriol.* **2005**, *187*, 6128–6136. [CrossRef]
29. Boekhorst, J.; Helmer, Q.; Kleerebezem, M.; Siezen, R.J. Comparative analysis of proteins with a mucus-binding domain found exclusively in lactic acid bacteria. *Microbiology* **2006**, *152*, 273–280. [CrossRef]
30. Cortes-Perez, N.G.; Azevedo, V.; Alcocer-González, J.M.; Rodriguez-Padilla, C.; Tamez-Guerra, R.S.; Corthier, G.; Gruss, A.; Langella, P.; Bermúdez-Humarán, L.G. Cell-surface display of E7 antigen from human papillomavirus type-16 in *Lactococcus lactis* and in *Lactobacillus plantarum* using a new cell-wall anchor from lactobacilli. *J. Drug Target* **2005**, *13*, 89–98. [CrossRef]
31. Kleerebezem, M.; Boekhorst, J.; van Kranenburg, R.; Molenaar, D.; Kuipers, O.P.; Leer, R.; Tarchini, R.; Peters, S.A.; Sandbrink, H.M.; Fiers, M.W.; et al. Complete genome sequence of *Lactobacillus plantarum* WCFS1. *Proc. Natl. Acad. Sci. USA* **2003**, *100*, 1990–1995. [CrossRef]

32. Roos, S.; Jonsson, H. A high-molecular-mass cell-surface protein from *Lactobacillus reuteri* 1063 adheres to mucus components. *Microbiology* **2002**, *148*, 433–442. [CrossRef] [PubMed]
33. Mathiesen, G.; Sveen, A.; Brurberg, M.B.; Fredriksen, L.; Axelsson, L.; Eijsink, V.G. Genome-wide analysis of signal peptide functionality in *Lactobacillus plantarum* WCFS1. *BMC Genom.* **2009**, *10*, 425. [CrossRef]
34. Mathiesen, G.; Sveen, A.; Piard, J.C.; Axelsson, L.; Eijsink, V.G. Heterologous protein secretion by *Lactobacillus plantarum* using homologous signal peptides. *J. Appl. Microbiol.* **2008**, *105*, 215–226. [CrossRef] [PubMed]
35. Karlskas, I.L.; Maudal, K.; Axelsson, L.; Rud, I.; Eijsink, V.G.H.; Mathiesen, G. Heterologous protein secretion in lactobacilli with modified pSIP vectors. *PLoS ONE* **2014**, *9*, e91125. [CrossRef] [PubMed]
36. Williamson, M.P. The structure and function of proline-rich regions in proteins. *Biochem. J.* **1994**, *297*, 249–260. [CrossRef]
37. Strych, U.; Penland, R.L.; Jimenez, M.; Krause, K.L.; Benedik, M.J. Characterization of the alanine racemases from two *Mycobacteria*. *FEMS Microbiol. Lett.* **2001**, *196*, 93–98. [CrossRef]
38. Aukrust, T.; Blom, H. Transformation of *Lactobacillus* strains used in meat and vegetable fermentations. *Food Res. Int.* **1992**, *25*, 253–261. [CrossRef]
39. Eijsink, V.G.; Brurberg, M.B.; Middelhoven, P.H.; Nes, I.F. Induction of bacteriocin production in *Lactobacillus sake* by a secreted peptide. *J. Bacteriol.* **1996**, *178*, 2232–2237. [CrossRef]
40. Bradford, M.M. A rapid and sensitive method for the quantitation of microgram quantities of protein utilizing the principle of protein-dye binding. *Anal. Biochem.* **1976**, *72*, 248–254. [CrossRef]
41. Rocha, R.; Almeida, C.; Azevedo, N.F. Influence of the fixation/permeabilization step on peptide nucleic acid fluorescence in situ hybridization (PNA-FISH) for the detection of bacteria. *PLoS ONE* **2018**, *13*, e0196522. [CrossRef]
42. Anzengruber, J.; Courtin, P.; Claes, I.J.J.; Debreczeny, M.; Hofbauer, S.; Obinger, C.; Chapot-Chartier, M.P.; Vanderleyden, J.; Messner, P.; Schäffer, C. Biochemical characterization of the major N-acetylmuramidase from *Lactobacillus buchneri*. *Microbiology* **2014**, *160*, 1807–1819. [CrossRef] [PubMed]
43. Nairn, R.C. *Fluorescent Protein Tracing*, 3rd ed.; E. & S. Livingstone Ltd.: Teviot Place, UK, 1969; p. 503.
44. Nguyen, H.A.; Nguyen, T.-H.; Nguyen, T.T.; Peterbauer, C.K.; Mathiesen, G.; Haltrich, D. Chitinase from *Bacillus licheniformis* DSM13: Expression in *Lactobacillus plantarum* WCFS1 and biochemical characterisation. *Protein Expr. Purif.* **2012**, *81*, 166–174. [CrossRef] [PubMed]

International Journal of
Molecular Sciences

MDPI

Article

Acceptor Specificity of β-*N*-Acetylhexosaminidase from *Talaromyces flavus*: A Rational Explanation

Cecilia Garcia-Oliva [1], Pilar Hoyos [1], Lucie Petrásková [2], Natalia Kulik [3], Helena Pelantová [2], Alfredo H. Cabanillas [4], Ángel Rumbero [4], Vladimír Křen [2], María J. Hernáiz [1,*] and Pavla Bojarová [2,5,*]

1 Department of Chemistry in Pharmaceutical Sciences, Faculty of Pharmacy, Complutense University of Madrid, Plaza Ramón y Cajal, E 28040 Madrid, Spain; cecilag@ucm.es (C.G.-O.); pilarhv@farm.ucm.es (P.H.)
2 Institute of Microbiology of the Czech Academy of Sciences, Vídeňská 1083, CZ 14220 Prague 4, Czech Republic; petraskova@biomed.cas.cz (L.P.); pelantova@biomed.cas.cz (H.P.); kren@biomed.cas.cz (V.K.)
3 Institute of Microbiology of the Czech Academy of Sciences, Center for Nanobiology and Structural Biology, Zámek 136, CZ 37333 Nové Hrady, Czech Republic; kulik@nh.cas.cz
4 Department of Organic Chemistry, Autonomous University of Madrid, Cantoblanco, 28049 Madrid, Spain; mjhernai@farm.ucm.es (A.H.C.); angel.rumbero@uam.es (Á.R.)
5 Department of Health Care Disciplines and Population Protection, Faculty of Biomedical Engineering, Czech Technical University in Prague, Nám. Sítná 3105, CZ 27201 Kladno, Czech Republic
* Correspondence: mjhernai@ucm.es (M.J.H.); bojarova@biomed.cas.cz (P.B.); Tel.: +34-913-941-820 (M.J.H.); +420-296-442-360 (P.B.)

Received: 18 November 2019; Accepted: 5 December 2019; Published: 7 December 2019

Abstract: Fungal β-*N*-acetylhexosaminidases, though hydrolytic enzymes in vivo, are useful tools in the preparation of oligosaccharides of biological interest. The β-*N*-acetylhexosaminidase from *Talaromyces flavus* is remarkable in terms of its synthetic potential, broad substrate specificity, and tolerance to substrate modifications. It can be heterologously produced in *Pichia pastoris* in a high yield. The mutation of the Tyr470 residue to histidine greatly enhances its transglycosylation capability. The aim of this work was to identify the structural requirements of this model β-*N*-acetylhexosaminidase for its transglycosylation acceptors and formulate a structure–activity relationship study. Enzymatic reactions were performed using an activated glycosyl donor, 4-nitrophenyl *N*-acetyl-β-D-glucosaminide or 4-nitrophenyl *N*-acetyl-β-D-galactosaminide, and a panel of glycosyl acceptors of varying structural features (*N*-acetylglucosamine, glucose, *N*-acetylgalactosamine, galactose, *N*-acetylmuramic acid, and glucuronic acid). The transglycosylation products were isolated and structurally characterized. The C-2 *N*-acetamido group in the acceptor molecule was found to be essential for recognition by the enzyme. The presence of the C-2 hydroxyl moiety strongly hindered the normal course of transglycosylation, yielding unique non-reducing disaccharides in a low yield. Moreover, whereas the *gluco*-configuration at C-4 steered the glycosylation into the β(1-4) position, the *galacto* acceptor afforded a β(1-6) glycosidic linkage. The Y470H mutant enzyme was tested with acceptors based on β-glycosides of uronic acid and *N*-acetylmuramic acid. With the latter acceptor, we were able to isolate and characterize one glycosylation product in a low yield. To our knowledge, this is the first example of enzymatic glycosylation of an *N*-acetylmuramic acid derivative. In order to explain these findings and predict enzyme behavior, a modeling study was accomplished that correlated with the acquired experimental data.

Keywords: β-*N*-acetylhexosaminidases; substrate specificity; transglycosylation; Glide docking; *Talaromyces flavus*; muramic acid; non-reducing carbohydrate

1. Introduction

The study of carbohydrates is essential to understand their role in the diverse biological activities in which they occur, mainly as part of glycoconjugates. Inflammation, coagulation, metastasis, immunomodulation, bacterial and viral infection are some of the processes which can be understood in depth through the synthesis of glycomimetics [1–4].

β-*N*-Acetylhexosaminidases (EC 3.2.1.52, CAZy GH20) are known in nature for their hydrolytic activity. They catalyze the cleavage of the non-reducing end of oligosaccharide chains containing *N*-acetyl-β-D-glucosaminide (GlcNAc) or *N*-acetyl-β-D-galactosaminide (GalNAc) units. However, in vitro, the reaction equilibrium can be shifted in favor of transglycosylation by changing the kinetic conditions (activated glycosyl donor, donor/acceptor ratio) and/ or by mutating the enzyme in the active site [5,6].

The broad range of substrates accepted makes these enzymes ideal synthetic tools with non-natural substrates [7]. In this context, β-*N*-acetylhexosaminidases from fungi have been demonstrated as attractive catalysts for this purpose [8–10]. Moreover, genetic engineering has enabled the modification of these enzymes, making them more convenient for the synthesis and expanding the spectrum of identified substrates [11].

The standard reaction used to test the transglycosylation performance of β-*N*-acetylhexosaminidases involves the use of an activated glycosyl donor such as *p*NP-GlcNAc or *p*NP-GalNAc and a glycosyl acceptor, for example, GlcNAc. The ratio between the donor and acceptor substrates must be low enough to promote the synthesis and minimize the hydrolysis. Chitooligosaccharides, the natural substrates of β-*N*-acetylhexosaminidases, consist of subunits linked by a β(1-4) glycosidic bond, which is thus the preferred linkage when oligosaccharide chains are formed [12]. β-*N*-Acetylhexosaminidases utilize a specific type of retaining catalytic mechanism of glycosidases, termed substrate-assisted catalysis. It is typical of the formation of an oxazolinium reaction intermediate (Supplementary Materials, Scheme S1).

In this work, we have focused on the β-*N*-acetylhexosaminidase from *Talaromyces flavus* CCF 2686 (*Tf*Hex), a representative of transglycosylating β-*N*-acetylhexosaminidases, due to its great substrate flexibility, which enhances its potential in synthetic reactions, and the availability of the molecular model of this enzyme. In addition, site-directed mutagenesis has allowed the improvement of its synthetic properties, generating a considerable increase in its transglycosylation capability through the substitution of the enzyme active site residue tyrosine 470 by histidine [11,13]. The *Tf*Hex Y470H mutant has been demonstrated as an efficient synthetic tool [14]. When quantifying its hydrolytic activity, the wild-type enzyme (*Tf*Hex WT) is ca. 200 times more efficient than *Tf*Hex Y470H. While *Tf*Hex WT is more active when hydrolyzing *p*NP-GalNAc, *Tf*Hex Y470H has a preference for the *pNP*-GlcNAc donor substrate.

So far, the acceptor specificity of these enzymes has not been thoroughly analyzed since previous studies have mainly focused on donor recognition [8,15]. Therefore, this work aims to explore the synthetic potential of *Tf*Hex WT and *Tf*Hex Y470H with a range of acceptor substrates, by gathering the structural requirements of the enzyme and exploring the spectrum of feasible reactions. In the frame of this study, a variety of natural monosaccharides derived from a *gluco*- or *galacto*-configuration were selected as acceptors: *N*-acetylglucosamine (GlcNAc; **1**), glucose (Glc; **2**), *N*-acetylmuramic acid (MurNAc; **3**), glucuronic acid (GlcA; **4**), galactose (Gal; **6**), and *N*-acetylgalactosamine (GalNAc; **5**), as well as other non-natural glycosides of *N*-acetylmuramic acid and glucuronic acid, functionalized at the anomeric position [16]. These compounds share a similar structure, but differ in the position or presence of some substituents. Through this strategy, we aimed to deduce which structural moieties in the acceptors with the *gluco*- or *galacto*-core are essential for the recognition by *Tf*Hex. A detailed molecular modeling study was performed to understand the enzyme behavior and determine the key residues participating in the interaction.

2. Results

2.1. Screening of the Glycosylation of Various Acceptors

2.1.1. Glycosylation Reaction Catalyzed by *Tf*Hex WT: Screening of Natural Monosaccharides

As exemplified above, *Tf*Hex WT is an excellent tool in carbohydrate synthesis [7]. For the aim of exploring the spectrum of acceptor substrates of this enzyme, we first decided to test its activity in the transglycosylation reaction involving *p*NP-GlcNAc or *p*NP-GalNAc as glycosyl donors and various natural monosaccharides of the *gluco-* and *galacto*-configuration as acceptors, namely GlcNAc (**1**), Glc (**2**), MurNAc (**3**), GlcA (**4**), Gal (**6**), and GalNAc (**5**) (Figure 1). As GlcNAc is the natural acceptor substrate of this enzyme, the transglycosylation reaction employing *p*NP-GlcNAc and GlcNAc (**1**) was used as a reference reaction in order to optimize the enzyme amount and concentration of donor and acceptor substrates. The donor/acceptor ratio of 1/6 (50 mM donor and 300 mM acceptor) was found as the optimum to shift the equilibrium towards the formation of the transglycosylation product. First, reactions with acceptor substrates **1–6** were conducted at an analytical scale (200 μL) and in case product formation was detected, reactions were scaled up to a 3 mL volume. The outcome of respective transglycosylation reactions is shown in Figures S1–S6 in the Supplementary Materials.

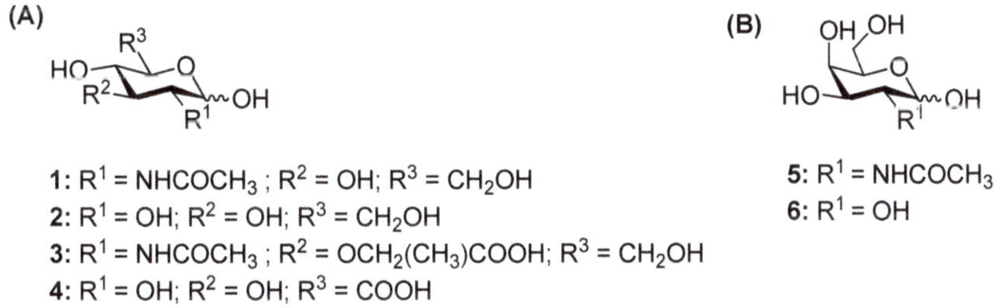

Figure 1. Selection of acceptors screened with the wild-type enzyme (*Tf*Hex WT). (**A**) Acceptors based on the *gluco*-configuration: *N*-acetylglucosamine (**1**), glucose (**2**), *N*-acetylmuramic acid (**3**), and glucuronic acid (**4**); (**B**) acceptors based on the *galacto*-configuration: galactose (**6**) and *N*-acetylgalactosamine (**5**).

The reference reaction between *p*NP-GlcNAc and GlcNAc (**1**) (Table 1, entry I), selectively afforded the corresponding *N,N′*-diacetylchitobiose product GlcNAc-β(1-4)-GlcNAc (**7**, Figure 2) in a 24% yield. In contrast, the reaction performed with the *p*NP-GalNAc donor and GalNAc (**5**) acceptor (Table 1, entry II), exclusively resulted in the β(1-6) regioisomer (**8**, Figure 2) in a 31% yield. As a rule of thumb, the axial *galacto*-position is considered practically impossible to glycosylate with β-*N*-acetylhexosaminidases [7]. This study demonstrates the high regioselectivity of β-*N*-acetylhexosaminidase-catalyzed glycosylation, depending on the configuration of the C-4 acceptor hydroxyl (Figure 2).

Table 1. Preparative transglycosylation reactions with *Tf*Hex WT and Y470H.

Entry	Enzyme	Donor	Acceptor	Donor/Acceptor [mM] [1]	Product	Yield
I	WT	*p*NP-GlcNAc	GlcNAc	50/300	GlcNAc-β(1-4)-GlcNAc	24%
II	WT	*p*NP-GalNAc	GalNAc	50/300	GalNAc- β (1-6)-GalNAc	31%
III	WT	*p*NP-GlcNAc	Glc	50/300	GlcNAc- β (1-1)-Glc	14%
IV	WT	*p*NP-GlcNAc	Gal	50/300	GlcNAc- β (1-1)-Gal	10%
V	Y470H	*p*NP-GlcNAc	MurNAc	50 (x2)/100	GlcNAc- β (1-X)-MurNAc	n.q. [1]
VI	Y470H	*p*NP-GalNAc	MurNAc-OPr	50/100	GalNAc- β (1-6)-MurNAc-OPr	≈1%

[1] Not quantified. The product could not be isolated in a sufficient purity for NMR characterization.

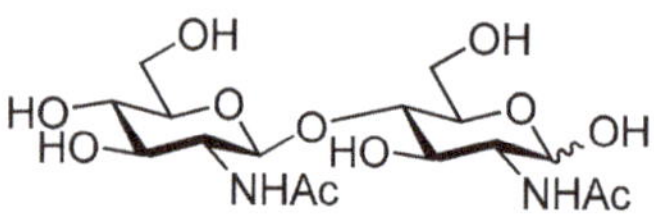

GlcNAc-β(1-4)-GlcNAc (**7**)

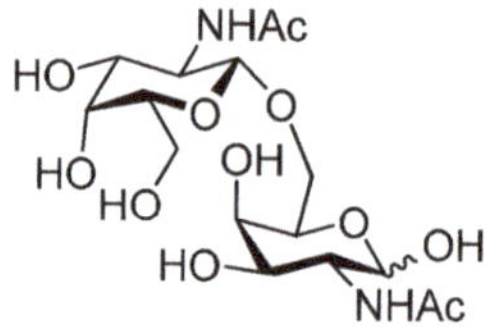

GalNAc-β(1-6)-GalNAc (**8**)

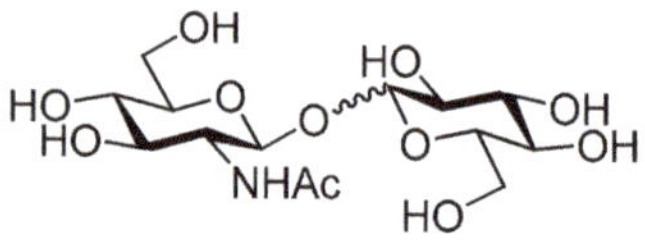

GlcNAc-β(1-1)-Glc (**9**)

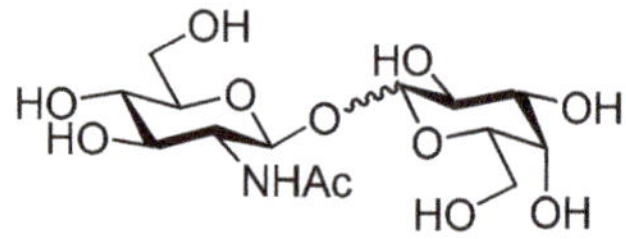

GlcNAc-β(1-1)-Gal (**10**)

Figure 2. Products of the glycosylation of GlcNAc (**1**), GalNAc (**5**), glucose (**2**), and galactose (**6**) catalyzed by *Tf*Hex WT.

When glucose (**2**) and galactose (**6**), lacking the C-2 *N*-acetamido group, were tested as acceptors (Table 1, entries III and IV, respectively), a small amount of the transglycosylation product was detected by thin layer chromatography (TLC) (Figures S3 and S4, Supplementary Material). After isolation by gel chromatography, the formation of a mixture of disaccharides was found by NMR spectroscopy, with the respective non-reducing β(1-1) disaccharide being among the major identified products (Figure 2, compounds **9** and **10**, respectively). The unexpected finding of non-reducing disaccharides as products of glycosylation of both glucose and galactose highlights the vital role of the *N*-acetamido group, not only for substrate recognition, but also for favorable acceptor orientation in the active site of the *Tf*Hex WT enzyme.

MurNAc (**3**) and GlcA (**4**) were also tested as challenging acceptors in the transglycosylation reactions mediated by *Tf*Hex WT. We speculated whether carboxy groups of MurNAc (**3**) and GlcA (**4**) may interfere in the accommodation of the compounds in the active site due to their negative charge. Therefore, in order to avoid the repulsion between the acceptor carboxy group and the charged residues in the enzyme's active site, methyl esters of **3** and **4** were first synthesized and tested as acceptors. However, the esters proved to be unstable under reaction and purification conditions, and were spontaneously hydrolyzed, affording the acidic form. Therefore, monosaccharides **3** and **4** with free carboxy groups were tested for glycosylation. In both cases, synthesis of the expected disaccharides failed and only the hydrolytic product, GlcNAc, could be detected, even when different fractions were analyzed after purification.

2.1.2. Glycosylation Reaction Catalyzed by TfHex Y470H: Screening of GlcA and MurNAc Glycosides

Since GlcA (**4**) and MurNAc (**3**) were not recognized as acceptors by the wild-type enzyme, they were subjected to testing using the mutant enzyme *Tf*Hex Y470H with suppressed hydrolytic activity and increased transglycosylation potential [13]. Besides free monosaccharides, a series of GlcA and MurNAc glycosides [16] functionalized at the anomeric position (**11**–**15**, Figure 3) were also employed as acceptors in the transglycosylation reaction mediated by this biocatalyst to test the influence of the anomeric effect on the glycosylation process. GlcA (**11a**–**15a**) and MurNAc glycosides (**11b**–**15b**) were prepared according to the protocol previously described [16].

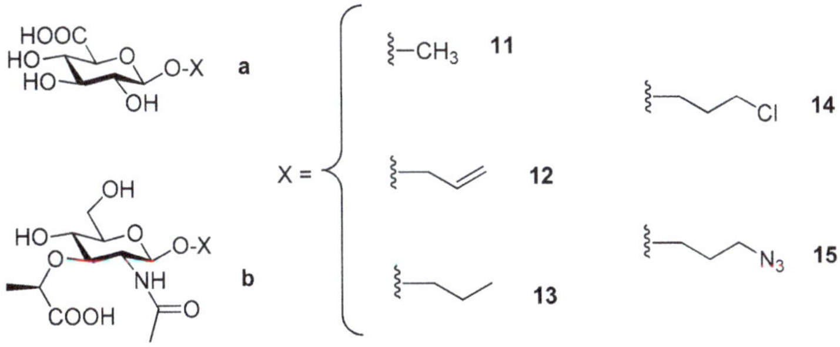

Figure 3. Series of β-glycosides of GlcA (**11a–15a**) and MurNAc (**11b–15b**) modified at the anomeric position.

When these acceptors were screened in transglycosylation reactions using *p*NP-GlcNAc as a donor, a range of chitooligomers of different lengths were detected in the reaction, which is typical of this mutant enzyme, as reported previously [11]. Therefore, it seemed that *p*NP-GalNAc may be the donor of preference since the formation of longer oligosaccharides was not observed with this mutant enzyme [14]. In accord with our previous experience, the optimum donor/acceptor ratio was higher thanks to the increased synthetic potential of the mutant enzyme and for the sake of minimizing the consumption of the acceptor employed. The most promising reactions, based on TLC and ESI-MS results, which were scaled-up, are summarized in Table 1. Unfortunately, no products were detected with either GlcA (**4**) or GlcA glycosides (**11a–15a**). In contrast, MurNAc (**3**) seemed to be better recognized by the enzyme. In the reaction with the *p*NP-GlcNAc donor and MurNAc acceptor, the expected product was detected by ESI-MS after 7 h. Regrettably, we failed to purify it to the extent to be analyzed by NMR in order to determine the regioselectivity of the reaction. In the screening of MurNAc glycosides (**11b–15b**), the best result was achieved when MurNAc-OPr (**13b**) was employed as an acceptor and *p*NP-GalNAc as a donor, affording the corresponding β(1-6) regioisomer **16** (Figure 4), though in a low yield (ca. 1%). We speculate that the presence of the lactate ether group at C-3 apparently impaired recognition by the enzyme, resulting in an unfavorable orientation for glycosylation. We further hypothesize that the presence of the bulky lactate ether promoted the glycosylation to the β(1-6) position, despite the otherwise strong enzyme selectivity for the β(1-4) bond in the case of GlcNAc. The low yield may also have been partially caused by the complex purification comprising gel permeation chromatography and silica gel chromatography.

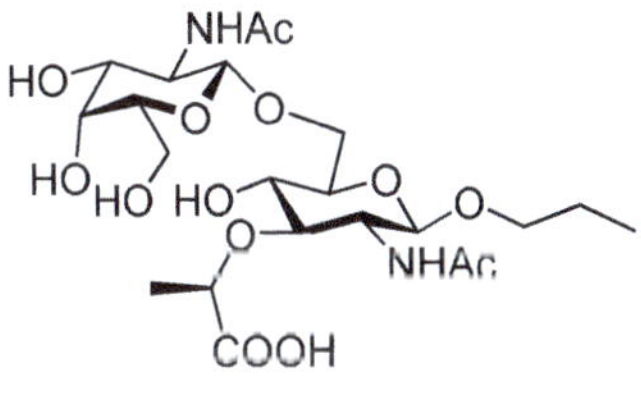

GalNAc-β(1-6)MurNAc-OPr (**16**)

Figure 4. Disaccharide **16** synthesized by the transglycosylation of MurA glycoside (MurNAc-OPr) **13b** with GalNAc under the catalysis by *Tf*Hex Y470H.

2.2. Docking and Molecular Dynamics

2.2.1. Docking and Molecular Dynamics Simulation of Selected Transglycosylation Acceptors in *Tf*Hex WT

To investigate the structural properties of *Tf*Hex and its ability to catalyze glycosidic bond formation, we performed the docking of transglycosylation acceptors in the active site of equilibrated complexes of WT and Tyr470His *Tf*Hex with GlcNAc-oxazoline (GlcNAcox) and GalNAc-oxazoline (GalNAcox) as mimics of intermediates during the hydrolysis of respective donors (*p*NP-GlcNAc or *p*NP-GalNAc). Interactions of GlcNAcox and GalNAcox intermediates with respective active site amino acid residues are shown in Figure S7 (Supplementary Materials). For equilibration parameters, see Figures S8 and S9 (Supplementary Materials). Scaled binding scores of transglycosylation acceptors are shown in Table 2.

Table 2. Scaled binding scores of docked transglycosylation acceptors (kcal/mol). The orientation of site 1 and site 2 is described in the text.

Acceptor	Position of Respective Acceptor Hydroxyl close to Catalytic Glu371						
	None (Site 1)	None (Site 2)	C-1	C-2	C-3	C-4	C-6
			***Tf*Hex WT in complex with GlcNAcox/GalNAcox**				
GlcNAc (**1**)	n.b.[1]/−3.99	−2.28/n.b.	n.b./n.b.	n.b./n.b.	−2.37/n.b.	−4.97/−3.36	−2.29/−2.09
Glc (**2**)	−5.24/−2.68	n.b./n.b.	-2.35/n.b.	-2.14/n.b.	−4.44/n.b.	−4.70/n.b.	−2.27/n.b.
MurNAc (**3**)	n.b./−4.46	−4.75/−2.93	n.b./n.b.	n.b./n.b.	n.b./n.b.	n.b./n.b.	−3.66/n.b.
GlcA (**4**)	−5.25/−5.47	−2.31/−2.60	n.b./n.b.	-2.03/n.b.	n.b./n.b.	−2.221/n.b.	n.b./n.b.
GalNAc (**5**)	−4.98/−2.62	−2.65/n.b.	n.b./n.b.	n.b./n.b.	−4.03/−2.28	n.b./n.b.	n.b./−2.23
Gal (**6**)	−2.57/−3.47	−2.36/n.b.	-2.29/n.b.	n.b./n.b.	−2.81/n.b.	n.b./n.b.	n.b./n.b.
MurNAc-*O*Pr (**XX**)	−2.42/−4.34	−4.41/−2.66	n.b./n.b.	n.b./n.b.	n.b./n.b.	n.b./n.b.	−4.12/−2.93
			***Tf*Hex Y470H in complex with GlcNAcox/GalNAcox**				
GlcA (**4**)	−5.45/3.80	n.b./−5.36	n.b./n.b.	n.b./n.b.	n.b./n.b.	n.b./n.b.	n.b./n.b.
MurNAc (**3**)	n.b./n.b.	−4.68/−4.61	n.b./n.b.	n.b./n.b.	n.b./n.b.	n.b./n.b.	n.b./−2.69
MurNAc-*O*Pr (**XX**)	−2.96/−3.19	−5.05/−4.50	n.b./n.b.	n.b./n.b.	n.b./n.b.	n.b./n.b.	n.b./−3.69

[1] n.b. stands for not bound. Two most favorable binding scores for each complex are indicated in red; the scores corresponding to bonds confirmed by the synthetic experiment are underlined.

There is an alternative binding place close to the active site with bound oxazoline intermediates, which is available for all acceptors—it is denoted here as site 1. In the WT enzyme, this site is more accessible for smaller ligands (Gal, Glc, GalNAc, GlcNAc) than for bulkier ones. Site 1 is placed close to the active site (Figure S10, Supplementary Materials) and is surrounded by residues Asn322, Glu332, Gly476, Gly477, Phe478, Arg484, Gly507, Trp509, Glu546, and Gln547, many of which belong to the flexible loops framing the active site [17]. This binding site is relatively well-accessible and is not a randomly formed temporary structure, as assumed from the ability of acceptors to dock in site 1 (Table 2) and from their stability in this position during molecular dynamics (MD) simulations (Figure S10, Supplementary Materials). However, acceptor binding in this site disables the productive outcome of the transglycosylation reaction due to a mutual unfavorable orientation of the donor and acceptor. Another alternative binding site—site 2—directs the binding of acceptors mostly above the donor, again with orientations unfavorable for the transglycosylation reaction— too far from the anomeric carbon of donor, too far from the catalytic base Glu371, or too far from both. This site is an unspecific binding site, the interactions of which greatly differ among individual acceptors. In Table 2, the affinity of both of these alternative binding sites to respective acceptors is presented with a binding score.

The binding score for the galactose acceptor in all found orientations with *Tf*Hex WT was the highest of all acceptors examined, so galactose may be considered the worst acceptor of the series. The most favorable binding scores were found for β(1-3) bond formation and for binding in site 1. To analyze the stability of docked galactose, we performed short molecular dynamics simulations, which revealed that in the case of β(1-3) binding, galactose is rotated unfavorably for product formation and the C-3 atom moves much further from the transglycosylation donor (Figure 5). Finally, during molecular dynamics with galactose, the catalytic Glu371 forms hydrogen bonds (HB) with C-3 and C-4 hydroxyls, which influences the charge distribution in the catalytic base Glu371 and complicates the involvement of respective hydroxyls in the formation of the glycosidic bond. Notably, in the case of galactose docked with C-1 close to Glu371, a close interaction with both Glu371 and the transglycosylation donor is preserved during a long molecular dynamics simulation time (Figure 5A). This could explain the enzyme preference for β(1-1) product formation. The ability of galactose to bind in site 1 with a similar score demonstrates a low specificity of binding and, together with the high value of the binding score, accounts for a low transglycosylation yield. Similarly, in the case of glucose, the binding in the WT-GlcNAcox system could lead to many products; the best binding scores are found for site 1, and for positions C-3 and C-4 close to catalytic Glu371. Orientation with the formation of the β(1–3) glycosidic bond is excluded for glucose as well as for galactose, as proved by molecular dynamics (Figure 5D) due to the acceptor rotation. In the *Tf*Hex WT-GalNAcox complex, galactose and glucose are not bound properly for productive transglycosylation due to the axial orientation of the C-4 hydroxyl of GalNAc oxazoline.

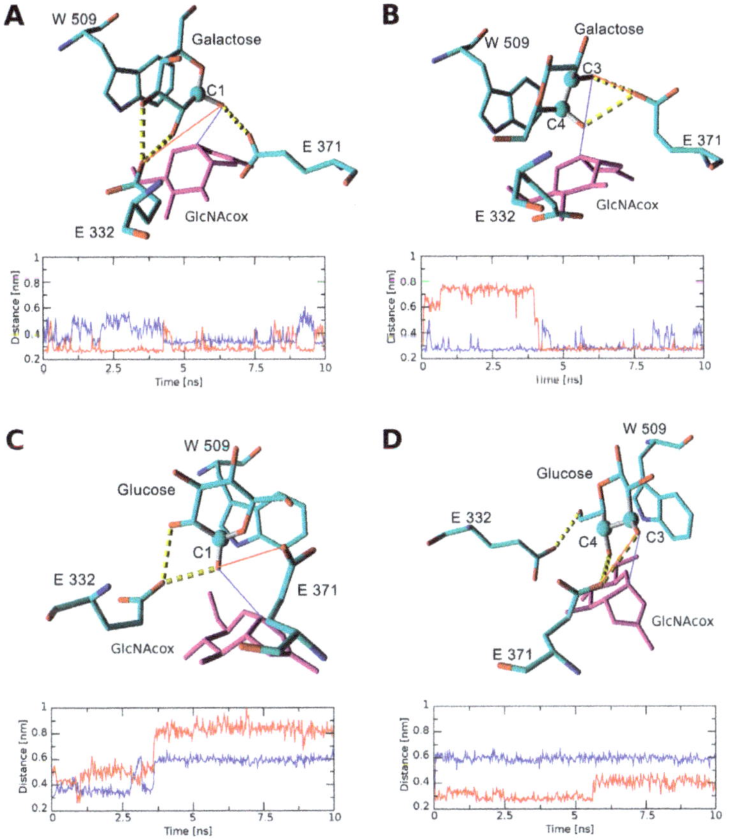

Figure 5. Interaction of glucose and galactose docked in the WT-GlcNAcox complex after 10 ns of molecular dynamics. Hydrogen bonds (HB) are shown by yellow dashed lines. Acceptor atoms docked close to Glu371 at the beginning of the molecular dynamics simulation are represented by a ball. (**A**) Snapshot of the WT-GlcNAcox complex with galactose docked with C-1 close to Glu371, and the distance between O-1 of Gal and the OE2 atom of Glu371 in (**red**) or C-1 of GlcNAcox (**blue**). The respective distances are indicated in the models in respective colors. (**B**) Snapshot of the WT-GlcNAcox complex with galactose docked with C-3 close to Glu371 and the distance between O-3 of Gal and the OE2 atom of Glu371 (**red**) or C-1 of GlcNAcox (**blue**). The respective distances are indicated in the models in respective colors. (**C**) Snapshot of the WT-GlcNAcox complex with glucose docked with C-1 close to Glu 371, and the distance between C-1 of GlcNAcox (**blue**) or the OE2 atom of Glu371 (**red**) and O-1 of glucose. The respective distances are indicated in the models in respective colors. (**D**) Snapshot of the WT-GlcNAcox complex with glucose docked with C-3 close to Glu371, and the distance between C-1 of GlcNAcox (**blue**) or the OE1 atom of Glu371 (**red**), and O-3 of glucose. The respective distances are indicated in the models in respective colors.

In the case of the GalNAc acceptor, the preferred orientation is with its C-3 hydroxyl close to the carboxyl of catalytic Glu371. However, the respective product with the β(1-3) bond cannot be formed due to the steric conflict between the *N*-acetyl group and Trp444 (Figure 6). Importantly, the inability to form the β(1-3) bond due to steric reasons may relate to all acceptors with *N*-acetyl substitution at C-2, namely GlcNAc, GalNAc, MurNAc, and MurNAc-*O*Pr. This inability is due to the unfavorable orientation of the *N*-acetyl group above hydrophobic Trp444. Concerning the GlcNAc acceptor docked with C-4 close to Glu371, its HB with Glu371, Glu332, and the backbone atoms of Val331 (interacting with the *N*-acetyl group) improves its stability in the active site. A favorable and stable interaction with Glu371 and GlcNAc oxazoline was found until 6.5 ns of MD. The binding scores with other possible orientations (C-3, C-6, and site 2) were twice as bad. This may account for a high yield and good selectivity of the β(1-4) product with GlcNAc (see also Figure S11, Supplementary Materials). GlcNAc can bind in a similar orientation to the WT-GalNAcox complex but with a lower productivity since it has a better binding score for site 1.

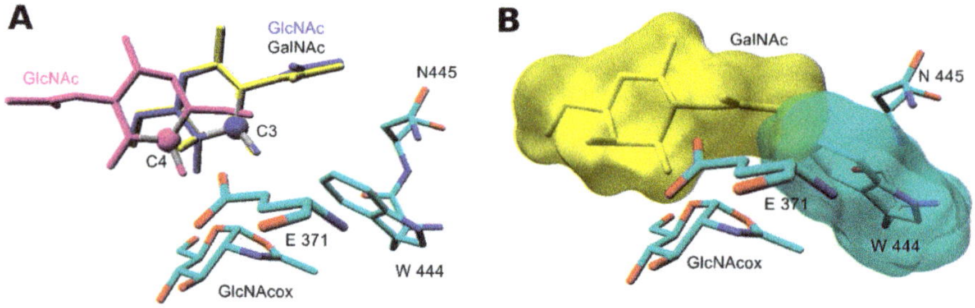

Figure 6. Steric conflict in the formation of the transglycosylation product with the β(1-3) glycosidic bond. (**A**). Overlay of docked orientations of GlcNAc (**blue**) and GalNAc (**yellow**) acceptors with C-3 or C-4 (GlcNAc in **magenta**) close to Glu371 in the WT-GlcNAcox complex. Active site residues within 3 Å from the GalNAc *N*-acetyl group are shown. (**A**) change that would be required in the C-3 position for product formation is indicated by the red dotted arrow. (**B**) Expected orientation of the GalNAc acceptor needed for the formation of GalNAc-β(1-3)-GlcNAc in the active site of WT-GlcNAcox. Position of the acceptor is determined from the alignment with *N,N'*-diacetylchitobiose (PDB ID: 1qbb). Intersection of the molecular surfaces of GalNAc (**yellow**) and Trp444 (**cyan**) shows that the GalNAc acceptor cannot move closer to the transglycosylation donor for the product formation due to steric hindrance.

With the GalNAc acceptor, the situation is similar. It binds with a lower productivity to the complex with GalNAc oxazoline (higher binding score for site 1), and the β(1-3) product formation is sterically hampered by Trp444 (Figure 6). MD simulation also confirmed a fast change of GalNAc acceptor orientation with water penetration close to Glu371 (data not shown). With both GalNAc and GlcNAc acceptors, the binding scores of β(1-6) bond formation are energetically less favorable. The orientation of acceptors with their C-6 close to C-1 of WT-GlcNAcox was unstable during molecular dynamics and resulted in a fast increase of this distance (data are not shown). However, due to the hindrance of β(1-3) bond formation, for the GalNAc acceptor, the β(1-6) bond is still the most feasible possibility as also demonstrated in the synthetic experiment.

The acceptors containing a carboxy group may exist in various protonation states; pK_a of GlcA varies from 2.83 to 3.28. Therefore, for the experimental situation at pH 5 used in this study, we modeled the respective deprotonated forms as prevalent under the transglycosylation conditions. In GlcA, the preferred orientation corresponds to acceptor binding in sites 1 or 2. In the case of MurNAc and MurNAc-*O*Pr in the WT-GlcNAcox complex, binding prevails at site 2, but C-6 is often close to Glu371 (Figure 7). Surprisingly, site 1 could not be targeted with a favorable score. There, a closer positioning of C-6 of MurNAc and MurNAc-*O*Pr was hindered by Glu332 rotation (Figure 7B). Another reason for the worse binding of MurNAc and MurNAc-*O*Pr of the WT-GlcNAcox complex is electrostatic—there are negatively charged residues in the vicinity of the acceptor carboxyl group when it binds closer to the transglycosylation donor (Figure 7 and Figure S12). The amino acids in the vicinity of the active site disallow C-6 hydroxyl to come closer to the transglycosylation donor and also restrict the conformational flexibility of MurNAc and MurNAc-*O*Pr acceptors. In the WT-GalNAcox complex, GlcA and MurNAc acceptors clearly bind best in the unproductive sites 1 or 2. Other orientations have more than 1 kcal/mol higher binding scores. During the molecular dynamics simulation of MurNAc in WT-GlcNAcox, the acceptor could approach GlcNAcox closer, but Glu371 was moved far from it and the active site was slightly destroyed. Additionally, due to the orientation of MurNAc, many water molecules come into the active site close to Glu371. In contrast, the MurNAc-*O*Pr acceptor maintained a favorable position in the active site during molecular dynamics and interactions with Glu371 and GlcNAcox were stable with low distances (data not shown). The docked orientation of MurNAc-*O*Pr in the Y470H mutant enzyme is slightly different (Figure S13), leading to a different transition state in the glycosylation step that may influence the product formation. Aliphatic propyl at C-1 may also reduce water access to the active site.

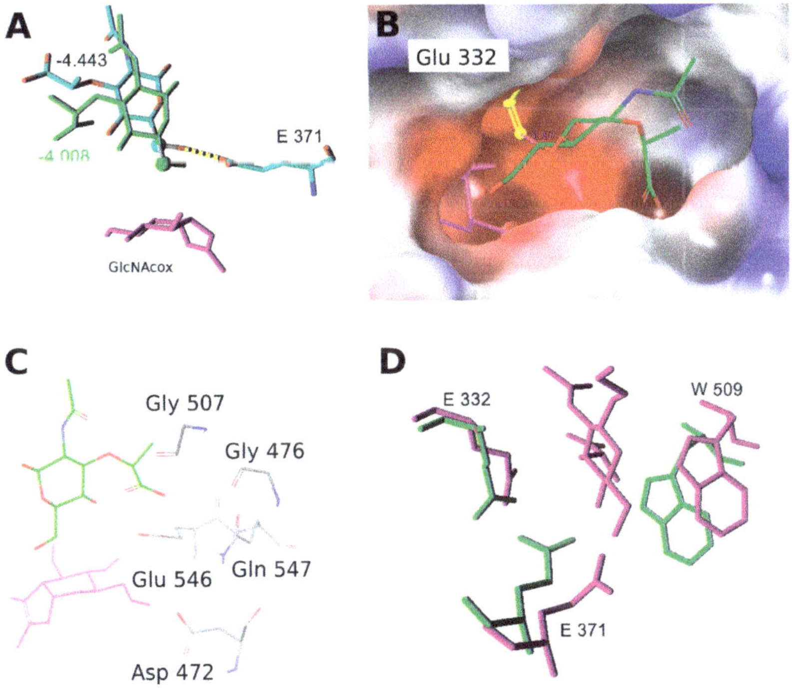

Figure 7. (**A**) Orientation of MurNAc in the WT-GlcNAcox complex: **green**—close to the transglycosylation donor; **element** color—far, corresponds to site 2. The respective Glide absolute binding scores are shown in the figure. GlcNAcox is shown in **magenta**, hydrogens are hidden, and the C-6 atom is represented by a ball. (**B**) Electrostatic potential surface of the WT enzyme active site with GlcNAcox (**magenta**) and the MurNAc acceptor (**element** color). Orientation of Glu332 (**yellow**) in one of the analyzed snapshots disallows a close interaction between the acceptor and donor. The negatively charged environment is shown in **red**, and that which is positively charged is presented in **blue**. (**C**) Residues with a negative charge in the vicinity of the MurNAc acceptor carboxyl group when it binds close to the transglycosylation donor. (**D**) Overlay of the aglycone binding site residues of WT-GlcNAcox (**green**) and Y470H-GalNAcox (**magenta**) with docked MurNAc-*O*Pr. Residues with similar side chain orientations are hidden.

2.2.2. Docking and Molecular Dynamics Simulation of Selected Transglycosylation Acceptors in *Tf*Hex Y470H

The mutant *Tf*Hex Y470H, previously demonstrated as an efficient synthetic tool [13,14], was selected for the study with difficult acceptors GlcA, MurNAc, and MurNAc-*O*Pr. All acceptors docked in the respective enzyme-oxazoline complexes showed unfavorable scores for binding, with the most favorable poses in sites 1 or 2. The Y470H-GalNAcox complex showed a low affinity to all studied acceptors, except for MurNAc-*O*Pr. We compared amino acid residues close to docked MurNAc-*O*Pr (Figure 7D) in the WT and mutant enzymes and found that residues Trp509 and Glu332 of the aglycone biding site have different side chain orientations, and thus influence the size of the aglycone binding site. The mutation of Tyr470 to His led to the loss of HB with the oxazoline intermediate and to the formation of different HB interactions with residues close to Trp509 (Supplementary Materials, Figure S7, and [13]). The Y470H mutant differs from the WT enzyme not only in the orientation of residues in the active site but also in the position of the oxazoline intermediate (Figure S7). The binding pocket surface, calculated for representative snapshots from molecular dynamics by Sitemap, showed a larger aglycone binding site surface for the Y470H-GalNAcox complex (1204.83 sq. Å) whereas for other combinations, it was significantly smaller (*cf.* 1003.468 sq. Å for WT-GalNAcox; 1094.485 sq. Å for Y470H-GlcNAcox; 839.991 sq. Å for WT-GlcNAcox).

3. Discussion

In this study, we aimed to thoroughly understand the performance of *Tf*Hex as a representative transglycosylating β-*N*-acetylhexosaminidase in terms of acceptor recognition. Therefore, we employed a series of acceptors with varying structural features based on *gluco*- or *galacto*-configuration. The enzyme tolerance to structural modifications in the donor molecule has been studied at great length [7]; however, little is known about the acceptor spectrum. *Tf*Hex has been known as an enzyme strictly selective for β(1-4) bond formation and a simplistic approach would suggest that if the activated donor is cleaved and an excess acceptor is present, disaccharide formation is nearly guaranteed. In this study, we demonstrate that the issue is far more complex and that the prediction of acceptor substrate specificity is even more challenging than with glycosyl donors. Docking allows us to identify possible binding poses of transglycosylation acceptors and reveal residues participating in the interaction. However, the number of expected products based on docking results is much higher than in the experiment. It is related to the uncertainty in the selection of an apt threshold for binders and of parameters for the validation of binding poses. Moreover, it appears that even a single point mutation in the active site may influence the enzyme behavior in far more aspects than supposed at first sight, as demonstrated in the example of the Y470H mutant.

Therefore, what structural aspects are critical for acceptor recognition and a good transglycosylation outcome? The first and foremost conclusion of this study is the vital importance of the C-2 *N*-acetamido moiety in the acceptor molecule. Whereas both GlcNAc and GalNAc acceptors are readily selectively glycosylated, the situation in the case of their structural counterparts Glc and Gal is drastically different. Not only is their glycosylation very poor, but it lacks selectivity and yields peculiar non-reducing disaccharides. This type of bond, where both reducing ends are implicated, has already been reported, to a lesser extent, during which non-natural substrates were employed with β-*N*-acetylhexosaminidases [18]. The theory highlighting the importance of the *N*-acetyl group at acceptor C-2 is further supported by the example of GlcA. Our previous experience showed that uronic acid of *p*NP-GlcNAc is readily hydrolyzed and subject to auto-condensation, forming the respective disaccharide [8]. However, with the GlcA acceptor void of an *N*-acetamido moiety, no disaccharide could be detected, irrespective of any substitution at C-1. Scaled docking binding scores allowed us to identify many possible orientations of transglycosylation acceptors, while further molecular dynamics refinement and analysis were needed to filter out false positive results.

The present results strongly support the rule of thumb on the inability of β-*N*-acetylhexosaminidases to glycosylate axial hydroxyls. Despite its high selectivity for the β(1-4) bond, the GalNAc acceptor was willingly glycosylated at C-6 with a high selectivity. In accordance with previous results, the β(1-3) bond was not detected in any case. As nicely explained by the docking studies, the formation of the β(1-3) linkage is hampered by the position of the substrate *N*-acetyl group in the enzyme active site.

The third main contribution of this study is the first ever example of the glycosylation of a challenging acceptor of MurNAc by a glycosidase, though with a low yield. Even here, the presence of a rather bulky substituent at C-3 has consequences for the reaction regioselectivity—in spite of the presence of the C-2 *N*-acetamido group, MurNAc is a very difficult acceptor, directing glycosylation regioselectivity to the β(1-6) position. Apparently, the presence of a larger substituent at C-1, such as propyl, enhanced the recognition of the substrate by the enzyme, being even more favorable in the Y470H mutant than in the WT. The seminal role of propyl may consist in reducing the access of water to the active site. This is because the different orientation of Glu332 and Trp509 regulates the size of the aglycone binding site and, together with GalNAcox, influences the mutant enzyme affinity to MurNAc-*O*Pr. Disaccharide GalNAc-β(1-6)-MurNAc-*O*Pr is the first enzymatically glycosylated MurNAc derivative and it opens new possibilities in enzymatic synthesis with modified compounds.

4. Materials and Methods

4.1. General Procedures

Commercial substrates, namely *p*NP-GlcNAc and *p*NP-GalNAc (Goldbio, St. Louis, CA, USA), GlcNAc (Acros Organics, Geel, Belgium), GalNAc (Glycon Biochemicals, Luckenwalde, Germany), D-glucose (Lach-ner, Neratovice, CZ), D-galactose, *N*-acetylmuramic acid, and D-glucuronic acid (all from Sigma-Aldrich, Munich, Germany), were employed without further purification. If not stated otherwise, other material was from Sigma-Aldrich, Munich, Germany. Thin layer chromatography (TLC) was performed on aluminium sheets precoated with Silica Gel 60 (F254 Merck, D), using 2-propanol/H_2O/NH_4OH aq. (7:2:1) as an eluent. Plates were visualized under UV light (254 nm) to detect the presence of UV active compounds and were then charred with a solution of 5% H_2SO_4 in ethanol.

4.2. Structural Analysis of Compounds

4.2.1. ESI-MS Analysis

Mass spectra were obtained using the Shimadzu Prominence system comprising a Shimadzu CBM-20A system controller, a Shimadzu LC-20AD binary HPLC pump, a Shimadzu CTO-10AS column oven, a Shimadzu SIL-20ACHT cooling autosampler, and a Shimadzu SPD-20MA diode array detector (Shimadzu, Tokyo, Japan). The ESI-MS parameters were as follows: positive and negative mode; ESI interface voltage 4.5 kV; detector voltage 1.15 kV; nebulizing gas flow 1.5 mL·min^{-1}; drying gas flow 15 mL·min^{-1}; heat block temperature 200 °C; temperature of desolvation line pipe 250 °C; SCAN mode 300–700 *m/z*. The mobile phase (acetonitrile) flow rate was 0.3 mL min^{-1}; chromatograms were analyzed using LabSolutions software ver. 5.75 SP2 (Shimadzu, Kyoto, Japan).

4.2.2. NMR Analysis

NMR spectra were acquired using a Bruker Avance III 600 MHz spectrometer (compounds **8**, **9**, and **10**, Bruker, Billica, MA, USA) or a Bruker Avance III 700 MHz spectrometer (compounds **7** and **16**) in D_2O (99.96 atom % D, VWR Chemicals, Leuven, Belgium) at 30 °C. The residual signal of water (δ_H 4.508) served as a reference in proton spectra; carbon spectra were referenced to the signal of acetone (δ_C 30.50). The structural assignment was based on information obtained from COSY, HSQC, HSQC-TOSCY, 1d-TOCSY, and HMBC experiments. The set of extracted vicinal $^3J_{H,H}$ coupling constants in individual sugar units allowed us to determine their anomeric configuration and to distinguish galactose and glucose moieties. The position of the glycosidic linkage was proved by HMBC contacts between involved carbons and anomeric protons of subsequent sugar units.

4.3. Enzymatic Activity Assay

β-*N*-Acetylhexosaminidase activity was determined spectrophotometrically (λ = 420 nm) in an end-point assay, by quantifying the release of 4-nitrophenol in the hydrolysis of *p*NP-GlcNAc and *p*NP-GalNAc (2 mM starting concentration) in McIlvaine (50 mM citrate-phosphate) buffer pH 5. Blank reactions were prepared analogously without an enzyme. Samples were incubated at 35 °C and 850 rpm for 10 min. The reaction must be stopped in time to afford only a low conversion of substrate (≤ ca 10%), in order to ensure the linear dependence of increasing *p*-nitrophenol formation on time. Then, the reactions were stopped by adding 1 mL of 0.1 M Na_2CO_3 and the amount of released 4-nitrophenol was determined colorimetrically. One unit of enzyme activity is defined as the amount of enzyme that cleaves 1 µmol of substrate per minute under the assay conditions.

4.4. Analytical Transglycosylation Reactions

The screening of transglycosylation reactions with the recombinant wild-type β-*N*-acetylhexosaminidase from *Talaromyces flavus* (*Tf*Hex WT) at an analytical scale (200 µL) was performed in McIlvaine buffer pH 5 in the presence of 50 mM donor and 300 mM acceptor (GlcNAc (**1**), Glc (**2**), MurNAc (**3**), GlcA (**4**), GalNAc (**5**), and Gal (**6**)). The glycosyl donor was *p*NP-GlcNAc, except when the GalNAc acted as the acceptor. *Tf*Hex WT was heterologously produced in *Pichia pastoris* and purified as described previously [13]. The final concentration of enzyme in the reaction mixture was 0.25 U mL^{-1}. The reaction mixtures were incubated at 35 °C under shaking at 1000 rpm. The reaction progress was monitored by TLC every 2 h up to 24 h or until all the donor had been consumed.

Analytical transglycosylation reactions with the *Tf*Hex Y470H mutant enzyme, where Tyr470 was substituted by histidine to suppress hydrolytic activity and promote transglycosylation, were performed similarly. The mutant enzyme was produced and purified as described previously [13]. The *p*NP-GalNAc donor (50 mM) and selected acceptor (GlcA, MurNAc or their glycosides) (100 mM) were suspended in 50 mM McIlvaine buffer pH 5 (final volume 200 µL). The reaction was started by the addition of *Tf*Hex Y470H (0.5 U mL^{-1} final volume) and it was incubated at 35 °C and 1000 rpm.

4.5. Preparative Transglycosylation Reactions

4.5.1. 2-Acetamido-2-Deoxy-β-D-Glucopyranosyl-(1-4)-2-Acetamido-2-Deoxy-D-Glucopyranose (GlcNAc-β(1-4)-GlcNAc; 7)

*p*NP-GlcNAc (3 × 17 mg, 50 mM) and the GlcNAc (**1**) acceptor (3 × 66 mg, 300 mM) were suspended in 3 × 1 mL of McIlvaine buffer pH 5 and the mixture was incubated at 35 °C and 1000 rpm. The reaction was started by adding *Tf*Hex WT (3 × 0.25 U; activity related to hydrolysis of *p*NP-GlcNAc). Reactions were monitored by TLC (Figure S1, Supplementary Materials) and stopped after 5 h by heating at 99 °C for 3 min. Reaction mixtures were cooled down to room temperature and centrifuged at 13,500 rpm for 10 min; the supernatant was loaded onto a Biogel-P2 column (15 × 900 mm) with water as a mobile phase. Upon analysis by TLC, fractions containing product **7** were combined and the remaining traces of *p*-nitrophenol were removed by extraction into ethyl acetate. After lyophilization, disaccharide 7 was obtained as a white solid (15 mg, 24% yield). ESI-MS: found *m/z* 521 for $[M + HSO_4]^-$, calcd for $C_{16}H_{29}N_2O_{15}S$ 521.1. ^1H and ^{13}C NMR data were consistent with the structure and were compared to the literature [19].

4.5.2. 2-Acetamido-2-Deoxy-β-D-Galactopyranosyl-(1-6)-2-Acetamido-2-Deoxy-D-Galactopyranose (GalNAc-β(1-6)-GalNAc; **8**)

*p*NP-GalNAc (3 × 17 mg, 50 mM) and the GalNAc (**5**) acceptor (3 × 66 mg, 300 mM) were suspended in 3 × 1 mL of McIlvaine buffer pH 5 and the mixture was incubated at 35 °C at 1000 rpm. The reaction was started by the addition of *Tf*Hex WT (3 × 0.25 U; activity related to hydrolysis of *p*NP-GalNAc). Reactions were monitored by TLC (Figure S2, Supplementary Materials) and stopped after 5.5 h by heating at 99 °C for 3 min. Reaction mixtures were cooled down to room temperature and centrifuged at 13,500 rpm for 10 min. Then, the supernatant was loaded onto a Biogel-P2 column (15 × 900 mm) with water as a mobile phase. Upon analysis by TLC, fractions containing product **8** were pooled together and the remaining traces of *p*-nitrophenol were removed by extraction into ethyl acetate. Lyophilization afforded product **8** as a white solid (19 mg, 31% yield). ESI-MS: found *m/z* 447 for $[M + Na]^+$, calcd for $C_{16}H_{28}N_2NaO_{11}$ 447.2; found *m/z* 521 for $[M + HSO_4]^-$, calcd for $C_{16}H_{29}N_2O_{15}S$ 521.1. The ^1H and ^{13}C NMR results are shown in the Supplementary Materials (Table S1, Figure S14a,b).

4.5.3. 2-Acetamido-2-Deoxy-β-D-Glucopyranosyl-(1-1)-β-D-Glucopyranoside (GlcNAc-β(1-1)-Glc; **9**)

The *p*NP-GlcNAc donor (3 × 17 mg, 50 mM) and Glc (**2**) acceptor (3 × 54 mg, 300 mM) were suspended in 3 × 1 mL of McIlvaine buffer pH 5 and the mixture was incubated at 35 °C at 1000 rpm. The reaction was started by the addition of *Tf*Hex WT (3 × 0.25 U; activity related to hydrolysis of *p*NP-GlcNAc). Reactions were monitored by TLC (Figure S3, Supplementary Materials) and stopped after 5.5 h when the donor was almost consumed by heating at 99 °C for 3 min. Reaction mixtures were cooled down to room temperature and centrifuged for 10 min at 13,500 rpm to remove the denatured enzyme. Then, the supernatant was loaded onto a Biogel-P2 column (15 × 900 mm) with water as a mobile phase. On the basis of TLC analysis, fractions containing product **9** were combined and extracted with ethyl acetate to remove traces of *p*-nitrophenol until a colorless solution was obtained. Since the lyophilized product fraction contained unwanted *p*-nitrophenyl *N,N'*-diacetylchitobioside (autocondensation product of *p*NP-GlcNAc) according to TLC, the second purification step was applied using an Amberlite XAD-2 (Sigma-Aldrich, Munich, Germany) column, eluted with water. The eluted fractions, void of aromatic impurities, were collected and lyophilized, to obtain impure non-reducing disaccharide **9** as a white solid (8 mg, approximate yield 14%). The sample of compound **9** contained other by-products, probably regioisomers, which could not be identified. ESI-MS: found *m/z* 384 for $[M + H]^+$, calcd for $C_{14}H_{26}NO_{11}$ 384.1; found *m/z* 406 for $[M + Na]^+$, calcd. 406.1 for $C_{14}H_{25}NNaO_{11}$; found *m/z* 418 for $[M + Cl]^-$, calcd. 418.1 for $C_{14}H_{25}ClNO_{11}$. ^1H and ^{13}C NMR data are shown in the Supplementary Materials (Table S2, Figure S15a,b).

4.5.4. 2-Acetamido-2-Deoxy-β-D-Glucopyranosyl-(1-1)-β-D-Galactopyranoside (GlcNAc-β(1-1)-Gal; **10**)

The preparative reaction with the Gal (**6**) acceptor (Figure S4, Supplementary Materials) was performed analogously to the Glc acceptor, only using Gal (3 × 54 mg, 300 mM). Purification was performed analogously to compound **9** and it afforded compound **10** as a white solid (6 mg, 10% yield). The sample of compound **10** contained other by-products, probably regioisomers, which could not be identified. ESI-MS: found *m/z* 384 for $[M + H]^+$, calcd for $C_{14}H_{26}NO_{11}$ 384.1; found *m/z* 406 for $[M + Na]^+$, calcd. 406.1 for $C_{14}H_{25}NNaO_{11}$; found *m/z* 418 for $[M + Cl]^-$, calcd. 418.1 for $C_{14}H_{25}ClNO_{11}$. The ^1H and ^{13}C NMR results are shown in the Supplementary Materials (Table S3, Figure S16a,b).

4.5.5. 2-Acetamido-2-Deoxy-β-D-Glucopyranosyl *N*-Acetylmuramic acid (GlcNAc-β(1-X)-MurNAc)

The *p*NP-GlcNAc donor (3 × 14 mg, 50 mM) and MurNAc (**3**) acceptor (3 × 24 mg, 100 mM) were suspended in 3 × 0.8 mL of McIlvaine buffer pH 5. After dissolving MurNAc, the pH of the reaction mixture was re-adjusted to pH 5 and the mixture was incubated at 35 °C at 1000 rpm. The reaction was started by the addition of *Tf*Hex Y470H (3 × 1.3 U; activity related to hydrolysis of *p*NP-GlcNAc).

Reactions were monitored by TLC (Figure S5, Supplementary Materials). After 3.5 h, another portion of donor was added (3 × 14 mg) and stopped after 7 h when the donor was almost consumed by heating at 99 °C for 3 min. Reaction mixtures were cooled down to room temperature and centrifuged for 10 min at 13,500 rpm to remove the denatured enzyme. Then, the supernatant was loaded onto a Biogel-P2 column (15 × 900 mm) with water as a mobile phase. TLC analysis showed that the desired product migrated together with the MurNAc acceptor. Fractions containing the product were combined and extracted with ethyl acetate to remove traces of *p*-nitrophenol. Lyophilization afforded partially purified product, as identified by ESI-MS analysis. ESI-MS: found *m/z* 497 for $[M + H]^+$, calcd for $C_{19}H_{33}N_2O_{13}$ 497.2; found *m/z* 495 for $[M - H]^-$, calcd for $C_{19}H_{31}N_2O_{13}$ 495.2. Unfortunately, due to a high concentration of the MurNAc acceptor in the partially purified product, NMR analysis could not be performed.

4.5.6. Propyl 2-Acetamido-2-Deoxy-β-D-Glucopyranosyl-(1-6)-*N*-Acetylmuramic acid (GalNAc-β(1-6)-MurNAc-*O*Pr; **16**)

*p*NP-GalNAc (3 × 17 mg, 50 mM) and the MurNAc-*O*Pr (**13b**) acceptor (3 × 34 mg, 100 mM) were suspended in 3 × 1 mL McIlvaine buffer pH 5 and the mixture was incubated at 35 °C at 1000 rpm. The reaction was started by the addition of *Tf*Hex Y470H (3 × 0.5 U, related to hydrolysis of *p*NP-GalNAc activity). Reactions were monitored by TLC (Figure S6, Supplementary Materials) and stopped after 3 h by heating at 99 °C for 3 min. Then, reaction mixtures were cooled down to room temperature and centrifuged at 13,500 rpm for 10 min. Following this, the supernatant was loaded onto a Biogel-P2 column (15 × 900 mm) with water as a mobile phase. Upon TLC analysis, fractions containing product **16** were combined and remaining traces of *p*-nitrophenol were extracted into ethyl acetate. After lyophilization, a second purification step, flash chromatography on silica gel, was conducted, using chloroform/methanol/$NH_4OH_{(aq)}$ in a ratio of 2:3:0.4 as mobile phase. Fractions were analyzed by TLC and those containing product **16** were pooled together and evaporated in vacuo to yield disaccharide **16** (2 mg, 1% yield). ESI-MS: found *m/z* 557 for $[M - H + H_2O]^+$, calcd for $C_{22}H_{41}N_2O_{14}$ 557.2. The ^1H and ^{13}C NMR results are shown in the Supplementary Materials (Table S4, Figure S17a,b).

4.6. Docking and Molecular Dynamics of the TfHex Active Site in the Complex with Tested Acceptors

The models used herein were adopted from those previously published [20]. The orientation of long loops and the *N*-terminal domain were corrected based on the crystal from *Aspergillus oryzae*; pdb: 5oar [17]. Molecular dynamics simulations were done for WT and Y470H mutants with docked GlcNAcox and GalNAcox ligands in the active site. Both oxazolines were docked into each monomer of the dimeric structure of *Tf*Hex and analyzed for 10 ns of molecular dynamics simulation, with the parameters explained in [13]. Several models with docked oxazolines were selected from the stable period of simulation (Figure S8, Supplementary Materials) for docking aglycones (each monomer was analyzed separately). To extend sampling of aglycone binding place, we analyzed several molecular dynamics snapshots with various Glu371, Trp509, and transglycosylation donor orientations, which could influence aglycone binding [21]. We defined two main parameters for the selection of enzyme–donor complexes from a production molecular dynamics run: distance from the catalytic Glu371 carboxy group and C-1 of oxazoline (Figure S9, Supplementary Materials), and the dihedral angle formed by CA-CB-CG-OE2 atoms in Glu 371. Structures with various dihedral angles and various distances were chosen, minimized with Yasara [22], and used for further docking experiments.

We performed the constraint and free docking with Glide [23]. Constraints for docking were as follows: distance between Glu371 (OE2 atom) and hydroxyl of the transglycosylation acceptor, and distance from hydroxyl of the acceptor to C-1 of the donor. The selection of these parameters was based on the analysis of the second step of hydrolysis/transglycosylation. During this step, two main events occur: donation of hydrogen to Glu371 from the acceptor/water with subsequent attack at C-1 of the donor [24]. An increase of these distances complicates the attack at donor C-1 and disallows the formation of the transglycosylation product [21]. Substrates were bound with a local docking algorithm with a grid, centered at Glu371 and Trp509, and extended in all directions by 10 Å from the center. Docked poses were scored with an SP Glide score and at least the five best structures were reported for each run. Docked poses with a proper geometry for transglycosylation and with the best binding scores were selected. Scores for structures with different Glu371 orientations were scaled based on the probability of occurrence, according to Equation (1):

$$\text{Scaled score} = P/n \sum_{i=1}^{n} SP_{Glide\ score}, \quad (1)$$

where P is the probability of observing a certain rotameric conformation of catalytic Glu371 during the production run, and n is the number of different orientations of catalytic residues. The change of Glu371 orientation with respect to the docked intermediate reflects the influence of the docking orientation of the transglycosylation acceptor and the docking score. To cover the possible catalytic residue conformations during molecular dynamics, we selected several representative structures with different Glu371 rotamers (Figure S9). For the molecular dynamics simulations with WT-GalNAcox and WT-GlcNAcox, the root mean square deviation of the compound was already stabilized after 3.5 ns; the period of 3.5–10 ns was used for calculation probabilities.

If a docking pose is found in all analyzed conformations, its score is below −4 kcal/mol. In case the mentioned orientation is unavailable in some conformations, the score is increased. As a result, for a good docking pose found in 50% MD of simulation snapshots, the scaled scores would be lower than −2 kcal/mol. This criterion was used for estimating a significant value to be reported.

The binding pocket surface was calculated by the Site map for the region within 3 Å, and extended to the distance of 3 Å to the solvent.

Refinement of the docked positions of transglycosylation acceptors was done by a molecular dynamics simulation (10–15 ns) with Yasara [22]. One repetition was conducted for each distinct orientation, with molecular dynamics parameters used for oxazolines.

5. Conclusions

The present work provides a rational explanation for the transglycosylation acceptor specificity of the *N*-acetylhexosaminidase from *Talaromyces flavus*. First, we found that the *N*-acetamido group is a crucial structural feature in the acceptor molecule. Both GlcNAc and GalNAc acceptors are selectively glycosylated with good yields. With the GalNAc acceptor, the enzyme exhibits a shift of its classic β(1-4) regioselectivity to β(1-6), which accords with the hypothesis that β-*N*-acetylhexosaminidases are unable to glycosylate the axial C-4 hydroxyl. In contrast, the respective structural counterparts Glc and Gal lacking the *N*-acetamido group yield a complex product mixture containing unique non-reducing disaccharides GlcNAc-β(1-1)-Glc or GlcNAc-β(1-1)-Gal, respectively, among the major products. Furthermore, it was found that tMurNAc, GlcA, and their β-glycosides are hardly or not tolerated as glycosylation acceptors by the WT enzyme. Fortunately, the Y470H mutant enzyme was able to recognize the MurNAc *O*Pr acceptor, affording the respective β(1-6) linked disaccharide with a low yield. Molecular modeling studies indicate that a long substituent at C-1, such as propyl, is beneficial for the acceptor specificity of *Tf*Hex. The GalNAc-β(1-6)-MurNAc-*O*Pr disaccharide is the first enzymatically glycosylated MurNAc derivative and it opens new possibilities in enzymatic synthesis with modified MurNAc derivatives.

Supplementary Materials: Supplementary Materials can be found at http://www.mdpi.com/1422-0067/20/24/6181/s1.

Author Contributions: Conceptualization, M.J.H., P.B., and V.K.; investigation, C.G.-O.; P.H.; L.P.; N.K., A.H.C., H.P., and Á.R.; data curation, L.P., N.K., and H.P.; writing—original draft preparation, P.B. and C.G.-O.; writing—review and editing, P.B., M.J.H., and V.K.; supervision, P.B.; funding acquisition, M.J.H., P.B., and V.K.

Funding: M.J.H. gratefully acknowledges financial support provided by the Spanish Ministerio de Economía y Competitividad, Grant RTI2018-096037-B-I00. P.B. and V.K. acknowledge support from LTC19038 (the Ministry of Education, Youth and Sports of the Czech Republic) and Czech Science Foundation project 18-01163S. N.K. acknowledges access to the computing facilities provided by the CESNET LM2015042 and the CERIT Scientific Cloud LM2015085 (as a part of the program "Projects of Large Research, Development, and Innovations Infrastructures").

Conflicts of Interest: The authors declare no conflicts of interest. The funders had no role in the design of the study; in the collection, analyses, or interpretation of data; in the writing of the manuscript; or in the decision to publish the results.

Abbreviations

ESI-MS	Electrospray ionization mass spectrometry
Gal	Galactose
GalNAc	*N*-Acetylgalactosamine
GalNAcox	*N*-Acetylgalactosamine oxazoline (catalytic intermediate during enzymatic hydrolysis)
Glc	Glucose
GlcA	Glucuronic acid
GlcNAc	*N*-Acetylglucosamine
GlcNAcox	*N*-Acetylglucosamine oxazoline (catalytic intermediate during enzymatic hydrolysis)
HB	Hydrogen bond
MurNAc	*N*-acetylmuramic acid
MurNAc-*O*Pr	Propyl glycoside of *N*-acetylmuramic acid
*p*NP-GalNAc	4-Nitrophenyl *N*-acetyl-β-D-galactosaminide
*p*NP-GlcNAc	4-Nitrophenyl *N*-acetyl-β-D-glucosaminide
*Tf*Hex WT	Wild type β-*N*-acetylhexosaminidase from *Talaromyces flavus*
*Tf*Hex Y470H	Mutant β-*N*-acetylhexosaminidase from *Talaromyces flavus*, Tyr470 is exchanged for His

References

1. Hakomori, S. Carbohydrate-to-carbohydrate interaction, through glycosynapse, as a basis of cell recognition and membrane organization. *Glycoconj. J.* **2004**, *21*, 125–137. [CrossRef] [PubMed]
2. Hevey, R. Strategies for the development of glycomimetic drug candidates. *Pharmaceuticals* **2019**, *12*, 55. [CrossRef] [PubMed]
3. Bourin, M.C.; Lindahl, U. Glycosaminoglycans and the regulation of blood coagulation. *Biochem. J.* **1993**, *289*, 313–330. [CrossRef] [PubMed]
4. Sun, L.; Middleton, D.R.; Wantuch, P.L.; Ozdilek, A.; Avci, F.Y. Carbohydrates as T-cell antigens with implications in health and disease. *Glycobiology* **2016**, *26*, 1029–1040. [CrossRef] [PubMed]
5. Slamová, K.; Bojarová, P. Engineered *N*-acetylhexosamine-active enzymes in glycoscience. *Biochim. Biophys. Acta Gen. Subj.* **2017**, *1861*, 2070–2087. [CrossRef] [PubMed]
6. Slamová, K.; Bojarová, P.; Petrásková, L.; Křen, V. β-*N*-Acetylhexosaminidase: What's in a name...? *Biotechnol. Adv.* **2010**, *28*, 682–693. [CrossRef] [PubMed]
7. Bojarová, P.; Bruthans, J.; Křen, V. β-*N*-Acetylhexosaminidases-the wizards of glycosylation. *Appl. Microbiol. Biotechnol.* **2019**, *103*, 7869–7881. [CrossRef]
8. Bojarová, P.; Slámová, K.; Křenek, K.; Gažák, R.; Kulik, N.; Ettrich, R.; Pelantová, H.; Kuzma, M.; Riva, S.; Adámek, D.; et al. Charged hexosaminides as new substrates for β-*N*-acetylhexosaminidase-catalyzed synthesis of immunomodulatory disaccharides. *Adv. Synth. Catal.* **2011**, *353*, 2409–2420. [CrossRef]
9. Loft, K.J.; Bojarová, P.; Slámová, K.; Křen, V.; Williams, S.J. Synthesis of sulfated glucosaminides for profiling substrate specificities of sulfatases and fungal β-*N*-acetylhexosaminidases. *ChemBioChem* **2009**, *10*, 565–576. [CrossRef]
10. Bojarová, P.; Křenek, K.; Kuzma, M.; Petrásková, L.; Bezouška, K.; Namdjou, D.-J.; Elling, L.; Křen, V. *N*-Acetylhexosamine triad in one molecule: Chemoenzymatic introduction of 2-acetamido-2-deoxy-β-D-galactopyranosyluronic acid residue into a complex oligosaccharide. *J. Mol. Catal. B Enzym.* **2008**, *50*, 69–73. [CrossRef]

11. Slámová, K.; Krejzová, J.; Marhol, P.; Kalachova, L.; Kulik, N.; Pelantová, H.; Cvačka, J.; Křen, V. Synthesis of derivatized chitooligomers using transglycosidases engineered from the fungal GH20 β-*N*-acetylhexosaminidase. *Adv. Synth. Catal.* **2015**, *357*, 1941–1950. [CrossRef]
12. Lodhi, G.; Kim, Y.-S.; Hwang, J.-W.; Kim, S.-K.; Jeon, Y.-J.; Je, J.-Y.; Ahn, C.-B.; Moon, S.-H.; Jeon, B.-T.; Park, P.-J. Chitooligosaccharide and its derivatives: Preparation and biological applications. *Biomed. Res. Int.* **2014**, *2014*, 654913. [CrossRef] [PubMed]
13. Bojarová, P.; Kulik, N.; Hovorková, M.; Slámová, K.; Pelantová, H.; Křen, V. The β-*N*-acetylhexosaminidase in the synthesis of bioactive glycans: Protein and reaction engineering. *Molecules* **2019**, *24*, 599. [CrossRef] [PubMed]
14. Bojarová, P.; Tavares, M.R.; Laaf, D.; Bumba, L.; Petrásková, L.; Konefal, R.; Bláhová, M.; Pelantová, H.; Elling, L.; Etrych, T.; et al. Biocompatible glyconanomaterials based on HPMA-copolymer for specific targeting of galectin-3. *J. Nanobiotechnol.* **2018**, *16*, 73. [CrossRef] [PubMed]
15. Slamová, K.; Gažák, R.; Bojarová, P.; Kulik, N.; Ettrich, R.; Pelantová, H.; Sedmera, P.; Křen, V. 4-Deoxy-substrates for β-*N*-acetylhexosaminidases: How to make use of their loose specificity. *Glycobiology* **2010**, *20*, 1002–1009. [CrossRef] [PubMed]
16. García-Oliva, C.; Cabanillas, A.H.; Perona, A.; Hoyos, P.; Rumbero, Á.; Hernáiz, M.J. Efficient synthesis of muramic and glucuronic acid glycodendrimers as dengue virus antagonist. *Chem. Eur. J.* **2019**. [CrossRef]
17. Škerlová, J.; Bláha, J.; Pachl, P.; Hofbauerová, K.; Kukačka, Z.; Man, P.; Pompach, P.; Novák, P.; Otwinowski, Z.; Brynda, J.; et al. Crystal structure of native β-*N*-acetylhexosaminidase isolated from *Aspergillus oryzae* sheds light onto its substrate specificity, high stability, and regulation by propeptide. *FEBS J.* **2018**, *285*, 580–598. [CrossRef]
18. Křen, V.; Rajnochová, E.; Huňková, Z.; Dvořáková, J.; Sedmera, P. Unusual nonreducing sugar GlcNAcβ(1-1)Manβ formation by β-*N*-acetylhexosaminidase from *Aspergillus oryzae*. *Tetrahedron Lett.* **1998**, *39*, 9777–9780. [CrossRef]
19. Bojarová, P.; Chytil, P.; Mikulová, B.; Bumba, L.; Konefał, R.; Pelantová, H.; Krejzová, J.; Slámová, K.; Petrásková, L.; Kotrchová, L.; et al. Glycan-decorated HPMA copolymers as high-affinity lectin ligands. *Polym. Chem.* **2017**, *8*, 2647–2658. [CrossRef]
20. Kulik, N.; Slámová, K.; Ettrich, R.; Křen, V. Computational study of β-*N*-acetylhexosaminidase from *Talaromyces flavus*, a glycosidase with high substrate flexibility. *BMC Bioinform.* **2015**, *16*, 28. [CrossRef]
21. Brás, N.F.; Fernandes, P.A.; Ramos, M.J. Docking and molecular dynamics studies on the stereoselectivity in the enzymatic synthesis of carbohydrates. *Theor. Chem. Acc.* **2009**, *122*, 283. [CrossRef]
22. Krieger, E.; Darden, T.; Nabuurs, S.B.; Finkelstein, A.; Vriend, G. Making optimal use of empirical energy functions: Force-field parameterization in crystal space. *Proteins* **2004**, *57*, 678–683. [CrossRef] [PubMed]
23. Friesner, R.A.; Banks, J.L.; Murphy, R.B.; Halgren, T.A.; Klicic, J.J.; Mainz, D.T.; Repasky, M.P.; Knoll, E.H.; Shelley, M.; Perry, J.K.; et al. Glide: A new approach for rapid, accurate docking and scoring. 1. Method and assessment of docking accuracy. *J. Med. Chem.* **2004**, *47*, 1739–1749. [CrossRef] [PubMed]
24. Mark, B.L.; Vocadlo, D.J.; Knapp, S.; Triggs-Raine, B.L.; Withers, S.G.; James, M.N. Crystallographic evidence for substrate-assisted catalysis in a bacterial β-hexosaminidase. *J. Biol. Chem.* **2001**, *276*, 10330–10337. [CrossRef]

International Journal of
Molecular Sciences

MDPI

Article

Structure-Based Redesign of a Self-Sufficient Flavin-Containing Monooxygenase towards Indigo Production

Nikola Lončar [1,†], Hugo L. van Beek [2,†] and Marco W. Fraaije [2,*]

1 GECCO Biotech, 9747 AG Groningen, The Netherlands; n.loncar@gecco-biotech.com
2 Molecular Enzymology group, University of Groningen, 9747 AG Groningen, The Netherlands; h.l.van.beek@rug.nl
* Correspondence: m.w.fraaije@rug.nl; Tel.: +31-50-3634345
† These authors contributed equally to this work.

Received: 31 October 2019; Accepted: 4 December 2019; Published: 5 December 2019

Abstract: Indigo is currently produced by a century-old petrochemical-based process, therefore it is highly attractive to develop a more environmentally benign and efficient biotechnological process to produce this timeless dye. Flavin-containing monooxygenases (FMOs) are able to oxidize a wide variety of substrates. In this paper we show that the bacterial mFMO can be adapted to improve its ability to convert indole into indigo. The improvement was achieved by a combination of computational and structure-inspired enzyme redesign. We showed that the thermostability and the k_{cat} for indole could be improved 1.5-fold by screening a relatively small number of enzyme mutants. This project not only resulted in an improved biocatalyst but also provided an improved understanding of the structural elements that determine the activity of mFMO and provides hints for further improvement of the monooxygenase as biocatalyst.

Keywords: indigo; MISO library; flavin; monooxygenase; FMO

1. Introduction

Indigo is one of the oldest dyes known to mankind. It was used in ancient Egypt to dye clothes as early as 2300 B.C. [1]. Originally indigo was produced from plants such as *Indigofera spp.* and *Polygonum tinctorum* [1]. The chemical process that was developed in 1883 by Adolph von Baeyer is still used in the 21st century [2]. One of the advantages of synthetic indigo is the purity, already reaching >90% in 1900 [3], while natural indigo from *Indigofera tinctoria* is obtained with a purity of between 20% and 90% [3]. Enzymatic oxidation of indole could result in indigo of a higher purity as enzymes are typically highly selective. The current annual production of indigo is estimated at 80,000 tons per year [4] which is mainly used as a dye in the textile industry to produce blue jeans, applying 3–12 g of indigo to dye a pair of jeans [4].

The demand for a cleaner process triggered microbiological research concerning indigo production. In 1983 the first naphthalene oxidation genes were expressed in *Escherichia coli* allowing indigo formation in the presence of tryptophan or indole [5], and which was further exploited [6]. Metabolic engineering was used to increase in vivo indole concentrations by modifying the tryptophan metabolism pathway combined with tuning expression of naphthalene dioxygenase (NDO) from *Pseudomonas putida* [7]. However, more extensive metabolic engineering is needed to adjust the overall metabolism of *E. coli* for indigo production [7]. Another disadvantage of NDOs is their susceptibility to inactivation [8]. In later studies it was demonstrated that various other enzymes can also be used for indigoid dye production, such as P450 monooxygenases [9,10] and styrene monooxygenases [11,12]. Another promising alternative enzyme was described by Choi et al. [13] who discovered the flavin-containing monooxygenase from *Methylophaga* sp. (mFMO) that is able to oxidize indole, resulting in the formation of indigo after non-enzymatic dimerization. Han et al. subsequently carried out optimization of indigo production by whole-cells containing mFMO yielding 920 mg/L of indigo [14]. Alfieri et al. (2008) solved the crystal structure of mFMO and described the moonlighting role of NADP in the structure of mFMO [15].

Wild-type mFMO is active towards indole but its substrate recognition and turnover rates are rather poor, especially when compared to other substrates such as trimethylamine [15] that show a k_{cat} that is more than eight-fold higher. This work aimed at improving the catalytic performance of the monooxygenase by structure-based enzyme redesign. To obtain an improved monooxygenase, first mutations were identified that improve the thermostability of mFMO using computational predictions [16,17]. Then, residues that form the substrate binding pocket have been replaced using site-directed mutagenesis. Mutations that led to improved catalytic efficiency were combined using multichange isothermal mutagenesis (MISO) [18] to create an mFMO mutant that is more effective in oxidizing indole. These experiments have resulted in an improved biocatalyst and provide more insight into the structural elements that tune the activity of mFMO. Together with more thorough characterization of this enzyme, this study provides a basis for further improvement of this monooxygenase as biocatalyst.

2. Results

Engineering a thermostable variant of mFMO—engineering an active site of the enzyme often results in mutants with decreased stability. In an ideal scenario, one would like to have an as stable as possible starting enzyme. mFMO falls into the category of moderately stable enzymes with an apparent melting temperature of 43.3 °C. The recently developed FRESCO protocol [16,17] is an effective tool for the stabilization of enzymes, including flavin-containing enzymes [19]. Through this computational protocol, mutations are predicted that should render the target protein more (thermo) stable. It involves the fully automated in silico generation of all possible single mutants and the respective folding energy calculations of the modeled structures of these mutants. In a next step, mutant structures that appear more stable according to these calculations are subjected to molecular dynamics simulations to filter out mutations that are not compatible with the structural dynamics. In a final visual inspection round, typically 100–200 mutations are selected that have the best energy scores. As a first step towards transforming mFMO into a robust biocatalyst, the FRESCO analysis was performed. Based on this, 140 mutant proteins were expressed and purified. Analysis of all these mutants resulted in the identification of 16 mutant enzymes that show an increase of apparent melting temperature of >1 °C (Table S1). Combining two mutations at the N-terminus (M15L and S23A) resulted in a 3 °C increase in apparent melting temperature. Both M15L and S23A are far from the active site (>17 Å) and we observed no significant effect on the kinetic parameters of the enzyme. Unfortunately, adding more stabilizing mutations did not contribute to a higher thermostability. In the rest of the manuscript, *mFMO is used to indicate the use of this M15L/S23A double mutant which is a more stable starting point for the protein engineering campaign aimed at improving its catalytic performance on indole.

Engineering mFMO for improved catalytic performance with indole—in order to improve the activity of mFMO on indole we focused on the active site of the enzyme. When inspecting the crystal structure of mFMO that contains both cofactors, FAD and $NADP^+$, a cavity close to the reactive moiety of the flavin cofactor can be identified [15]. The residues that form the entrance and shape of this cavity are highlighted in Figure 1 and listed in Table 1). Several mutant enzymes targeting the active site residues and combinations thereof have been expressed, purified, and evaluated using kinetic measurements.

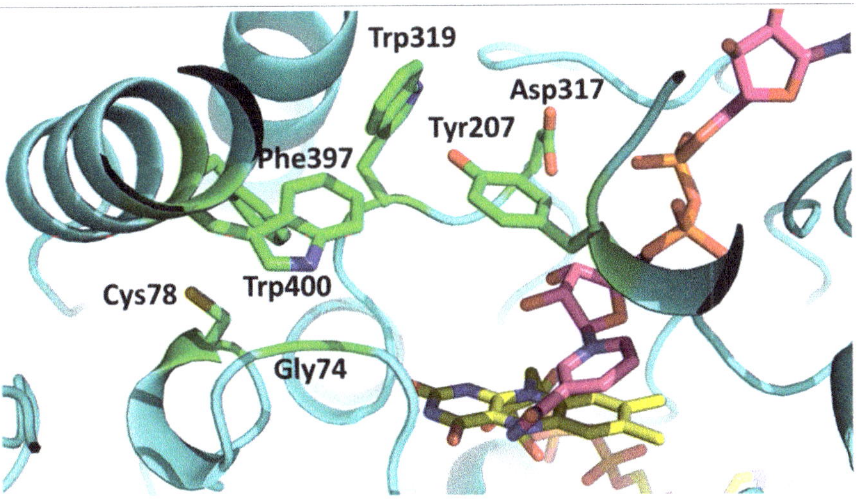

Figure 1. Active site of flavin-containing monooxygenase from *Methylophaga* sp. (mFMO). The residues forming a pocket next to the reactive moiety of the flavin cofactor are labeled and highlighted in green. The FAD cofactor is in yellow and the bound $NADP^+$ is in magenta. (PDB:2VQ7).

Table 1. Amino acid residues expected to affect substrate binding in mFMO.

Position	Comments
Gly74	A sidechain at this position could have interactions with the substrate. Despite being a pre-Pro residue, the phi-psi angles (−130,175°) will also fit non-Glycine residues.
Cys78	Its mutation could fill up a cavity in the active site making binding of indole more productive.
Tyr207	Forms part of entrance to the substrate binding cavity.
Asp317	Forms part of entrance to the substrate binding cavity.
Trp319	Limits the size of the substrate binding cavity.
Phe397	Limits the size of the substrate binding cavity.
Trp400	Limits the size of the substrate binding cavity.

In this first set of mutants we observed that they all retained activity (Table 2). Mutations on two positions resulted in beneficial effects: C78I and C78V increased the k_{cat} and C78A and Y207W decreased the K_M (although these mutations also decreased the k_{cat}). While mutating Trp319 did not directly lead to improvements, we remained interested in this position because it seems to have a major role in substrate binding. Based on these initial results we designed a MISO library. MISO allows the generation of a mutant library in which specific residues are allowed to mutate into a pre-designed set of alternative residues. Because mFMO activity could only be measured reliably using purified enzyme, we limited the size of the library to 48 possible mutations. For Cys78, we allowed the wild-type residue cysteine, the beneficial mutations to isoleucine, valine, and phenylalanine. For Tyr207, we allowed the wild-type residue tyrosine, the tryptophan, that was found to increase the affinity for indole, and asparagine. For Trp319, we allowed the wild-type tryptophan and phenylalanine, alanine, and asparagine. Instead of over screening by purifying (many) more mutants than the library size to get good coverage, we sequenced 190 clones first. In this set of clones, we obtained 44 of the 48 expected mutants. The respective mutant enzymes were obtained by growing cells at 15 mL scale and purification using 96 well plates [16]. After determining the concentration based on the absorbance at 441 nm, we performed a kinetic analysis of the mutants. We measured the uncoupling rate (the rate of NADPH consumption in the absence of substrate) and the activity with indole or TMA. A complete overview of the screened activities with indole and TMA and uncoupling rates can be found in the Supplementary Materials (Table S2).

Table 2. Kinetic parameters determined for *mFMO and mutants obtained from initial rational design and the multichange isothermal mutagenesis (MISO) library.

	Indole	
Variant	**k_{cat} (s^{-1})**	**K_M (mM)**
WT	0.85	0.4
	Site-directed mutants	
C78I	1.28	0.8
C78V	1.04	0.7
C78L	0.70	0.8
C78A	0.79	0.4
W319A	0.42	0.9
W319F	0.64	0.4
Y207W	0.31	0.1
	MISO mutants	
Y207W/W319A	0.80	0.8
C78I/Y207W/W319A	0.93	0.8

Screening of the MISO library led to the discovery of several mutants with increased activity compared to *mFMO (Table 2). The C78I mutant was the most active variant in the library, and also the C78V and Y207W/W319A mutants were identified as improved variants. Several other patterns were observed in the data obtained from screening the MISO library. Asparagine at position 207 leads to enzymes with a high uncoupling rate, still consuming NADPH but producing hydrogen peroxide instead of a hydroxylated product. A phenylalanine on position 78 leads to low activity. Several mutants, also some that only showed rates similar to *mFMO, were characterized in more detail (Table 2). Besides mutants showing a clear improvement in k_{cat} (C78I, C78V, C78I/Y207W/W319A) we also found mutants containing both the Y207W and W319A mutation to be active. These variants show behavior similar to the *mFMO when both mutations are introduced together. The Y207W mutation by itself reduces activity (but improved K_M) and the W319A by itself has a negative effect on both k_{cat} and K_M. This effect was observed in combination with the wild-type Cys78, as well as in combination with C78I and C78V.

Exploring conditions for mFMO-catalyzed indigo production—in addition to the engineering of the enzyme, we determined the optimal conditions for the enzyme. The pH optimum for activity was determined to be at pH 7.0–8.0, with most of the activity retained at pH 9 (Figure 2). The activity drops significantly at both lower and higher pH values. Interestingly, the stability of *mFMO shows a sharper pH optimum, with the highest thermostability at pH 8 (Figure 2). While at pH 9 the thermostability is still relatively high, it quickly drops at pH < 7 and pH > 9.

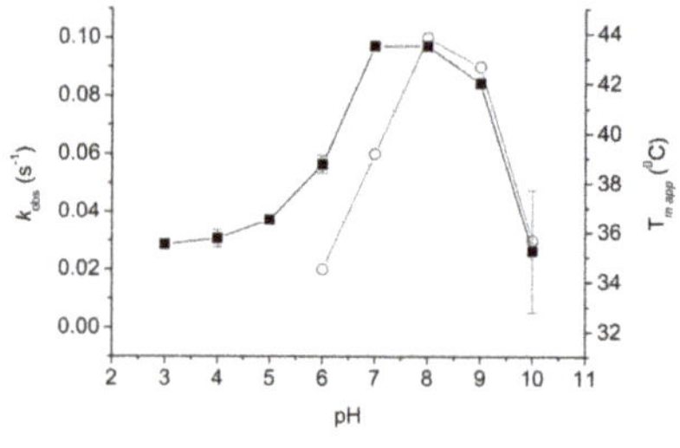

Figure 2. The effect of pH on the activity and stability of PTDH-*mFMO. (○)-activity (s^{-1}), (■)-stability as melting temperatures (°C).

The enzyme is optimally active at 250 mM NaCl, with an activity increase of more than 25% compared to the enzyme in the absence of NaCl (Figure S1). The enzyme is more stable with higher concentrations of NaCl. Glycerol has a slight stabilizing effect but no effect on activity (Figure S2A). DMSO up to 4% halves the activity while the enzyme remains quite stable (Figures S2B and S3B). Methanol also has a detrimental effect on stability, and dioxane is poorly tolerated by the enzyme (Figure S3B). Interestingly, $NADP^+$ as additive revealed an optimum for thermostability. In the range of 10–50 μM $NADP^+$, *mFMO is slightly stabilized, while higher concentrations lead to a lower thermostability (Figure S3A).

A layer of indigo-paste has to be scraped of the cell pellet before further processing and purification of *mFMO. Conversions with purified enzyme and cell-free extracts were done to quantify the indigo yield, as well as the purity with regards to the contaminant indirubin. The stability and activity measurements showed pH 8 is the optimum for performing conversions. NaCl and glycerol seem to have both a stabilizing effect as well as a positive effect on enzymatic activity. However, with the aim of developing a cost-effective process, no additives were used in the conversions. A reaction mixture containing only the essential components (see materials and methods) was used. To reduce the costs related to the use of NADPH, a catalytic amount of $NADP^+$ is used in combination with the fused cofactor regeneration system PTDH. Conversions were performed using either purified PTDH-*mFMO or PTDH-*mFMO cell free extract (CFE) (the use of CFE would reduce the cost of a process). This resulted in significant differences in both indigo yield and purity (Table 3). Indigo produced using CFE was found to be 94% pure while the use of purified enzyme resulted in >99% pure indigo. Synthetic indigo is known to reach purities of over 90% [20], with most available synthetic produced indigo having a purity of 94%. Purified PTDH-*mFMO seems to reach even higher purity levels, since the amount of indirubin is not detectable when measured with HPLC. We set up conversions in a such a way that the amount of enzyme would be limiting and a conversion of only 23% or 14%–18% was found for the purified enzyme or the CFE, respectively. The total turnover number was determined to be 5700, showing that there is some inactivation of the enzyme during the 24 h reaction time.

Table 3. Indigo yield after 24 h conversion of 10 mM indole with purified enzyme or two different batches of cell free extract (CFE).

	Indigo Yield (g/L)	Conversion (%)	Purity (%)	TTN
PTDH-mFMO	0.30	23	>99	5700
PTDH-mFMO CFE 1	0.24	18	94	n.a.
PTDH-mFMO CFE 2	0.18	14	94	n.a.

Conversion is calculated as the recovered amount of indigo compared to the theoretical yield. Indigo purity is reported as the percentage of indigo of the total amount of indigo and indirubin as analyzed by HPLC. The total turnover number (TTN) for CFE could not be determined as the exact amount of enzyme is not known. n.a.: not applicable.

3. Discussion

Already the wild-type mFMO expressing *E. coli* cultures develop an intense blue color clearly demonstrating the potential for indigo production. When the enzyme was studied in more detail, it was obvious that the enzyme prefers other substrates, for example trimethylamine over indole. This leaves room to improve the performance of mFMO on indole by engineering a variant with a higher affinity and catalytic rate by optimizing the fit of this molecule in the active site. In this study, we show that it is possible to improve the stability and catalytic performance on indole of this monooxygenase. The higher thermostability was achieved using the FRESCO protocol, which is a structure-based computational methodology which predicts mutations that improve thermostability of enzymes. Based on the structure of the enzyme we also generated and tested several mutants for improved catalytic performance. Small effects are expected from the second shell mutations, so the rather large effect of C78I is surprising. Based on the initial results with single mutants, a MISO library was built to further explore a small part of the sequence space. By first sequencing the library and subsequently only purifying unique mutants we were able to efficiently gather information about almost every mutant in the library. Unfortunately, this did not lead to mutants that are faster than the initially created C78I. By screening almost the entire library we obtained valuable knowledge on the mutability of the mFMO active site; we thoroughly probed Cys78, a very influential secondary-shell residue and found a synergistic mutation where a tryptophan is swapped with another aromatic residue. We also identified a mutation, Y207W, that results in a higher affinity for indole while also affecting the activity. The effect of the Y207W mutation on the catalytic performance of mFMO is in line with a previously reported Y207S mFMO mutant [21]. Cho et al. have shown that this specific mutation nearly abolishes all indole oxygenation activity of mFMO. Inspection of the available structures of mFMO suggests that Tyr207 is important in positioning the nicotinamide cofactor. Previous work has shown that mFMO and structurally related flavoprotein monooxygenases require a tightly bound nicotinamide cofactor, even after it has transferred a hydride to the flavin cofactor, for catalyzing oxygenations [22–26]. Thus, proper positioning of the nicotinamide cofactor in the $NADP^+$-complexed mFMO is important for indole oxidation. Tyr207 also points towards a cavity that can accommodate substrates in the $NADP^+$-bound mFMO structure. As a consequence, replacing Y207 can affect the activity and affinity for indole indirectly, through altering the position of the nicotinamide cofactor, or by directly influencing the substrate binding pocket.

We showed that the activity of *mFMO can be improved. The comprehensive results from the MISO library screening give a solid basis on which to design larger libraries. Some mutations primarily increased the k_{cat}, which is beneficial for any kind of indole conversion process, but depending on the used concentration of indole a lower K_M, which was also found for some mutants, might be equally desirable.

The thermostable *mFMO variant was used for conversions, both as purified enzyme and as cell-free extract. The relatively low conversion is primarily caused by the low enzyme loading: The measured total turnover number, 5700, could even be improved by applying a mutant enzyme or better conditions.

Conditions such as pH, solvents and salt concentrations were explored for the *mFMO. The optimum pH was eight or nine, with the activity decreasing more quickly than stability at lower pH. Solvents in general are detrimental for the enzyme, precluding cosolvent use.

4. Materials and Methods

4.1. Chemicals and Reagents

Indigo and indole were purchased from Sigma Aldrich (St. Louis, MO, USA). $NADP^+$ and NADPH were purchased from Oriental yeast Co (Tokyo, Japan). Other chemicals were analytical grade and obtained from either Sigma-Aldrich or Merck (Burlington, MA, USA).

4.2. Strains, Plasmids and Growth Conditions

Recombinant PTDH-mFMO was overexpressed and purified following the previously described procedure [27]. Briefly, *E. coli* NEB10β cells containing the pCRE3-mFMO plasmid grown overnight were diluted 100-fold in TBamp medium and induced directly with 0.02% arabinose. After incubation at 24 °C and 135 rpm (5 cm amplitude) for 48 h the cells were harvested (4 °C, 4000 × *g* for 15 min). The cell pellet was resuspended in 50 mM KPi buffer pH 7.5, disrupted by sonication, and the cell free extract was obtained by centrifugation at 4 °C 18,500 × *g* for 45 min. CFE was stored in 50 mM KPi buffer pH 8 containing 10% glycerol at −20°C after being frozen in liquid nitrogen. Purification of PTDH-mFMO was carried out on 5 mL HiTrap Ni-Sepharose column using ÄKTA system (Chicago, IL, USA). Concentrations of purified PTDH-mFMO was determined by using extinction coefficient of 14.4 mM^{-1} cm^{-1} at 441 nm.

4.3. Steady-State Kinetic Analyses

Kinetic assays were performed in triplicate using a BioTek synergy MX (Winooski, VT, USA) or BioTek synergy H1 platereader. Enzyme was used at 0.5 or 0.05 μM depending on the substrate. Different concentrations of substrate were used, up to 1.6 mM indole and 0.20 mM TMA. For indole, the solubility of the substrate was a limiting factor, for TMA the K_M values require a lower concentration of the substrate and a lower concentration of enzyme. Reactions were started by adding 100 μL NADPH to 100 μL enzyme-substrate mix. The depletion of NADPH was followed over time by measuring the 340 nm absorbance, using an extinction coefficient for NADPH of 6.22 mM^{-1} cm^{-1}. No correction was made based on the uncoupling rate in the absence of substrate, because the k_{cat} and K_M values were not significantly affected by this correction. The standard deviations for the determined kinetic parameters were typically <20%.

4.4. pH Optimum

Determination of the pH optimum of PTDH-mFMO activity was performed by using abovementioned procedure with the difference of using Britton Robinson buffer in pH range 2.0–12.0. Final concentrations in the 1 mL reaction mixture were 4 mM indole, 0.4 μM PTDH-mFMO and 0.2 mM NADPH.

4.5. ThermoFAD

The ThermoFAD method [28] was used to determine the apparent melting points of the PTDH-mFMO in different pH conditions or in presence of additives. Using an RT-PCR machine (CFX96-Touch, Bio-Rad, Hercules, CA, USA) the fluorescence of the FAD cofactor was monitored using a 450–490 excitation filter and a 515–530 nm emission filter, typically used for SYBR Green based RT-PCR. The temperature was increased with 0.5 °C per step, starting at 25 °C and ending at 90 °C, using a 10 s holding time at each step. The maximum of the first derivative of the observed flavin fluorescence was taken as the apparent melting temperature. 20 μL of the enzyme solution contain 10 μM PTDH-mFMO and the additive (solvent or buffer of appropriate pH) was put into a 96 wells PCR plate and covered with transparent cover, then ThermoFAD was ran using the BioRad CFX manager. The used additives were: Glycerol in concentrations of 0%, 1%, 5%, 10%, and 20%; NaCl in concentrations of 0, 50, 135, 250, and 500 mM; and $NADP^+$ in concentrations of 0, 1, 10, 50, and 100 μM. The used solvents were DMSO, methanol, and 1,4-dioxane all of which were used in concentrations of 0%, 1%, 5%, 10%, and 20%.

4.6. Indole Conversion by Purified PTDH-mFMO and Cell Free Extract (CFE)

Reactions of PTDH-mFMO with indole were set up to determine the efficiency of the indigo production. For this, final concentrations of 0.15 mM $NADP^+$, 0.4 μM FAD, 0.4 μM PTDH-mFMO, 20 mM Na-phosphite, 10 μL catalase (300 U), and 10 mM indole were used in 1 L 50 mM KPi buffer pH 8.0 in 2 L Erlenmeyer flask incubated at room temperature and stirred with magnetic stirrer. After 24 h the reaction mixture was centrifuged at 4000 × g to pellet the formed indigo. The indigo pellets were washed twice in demineralized water and once in 70% ethanol and dried until constant weight. Samples were dissolved in DMSO, diluted in methanol, and analyzed by HPLC. Conversions (1 L) were also performed with cell free extract containing PTDH-mFMO in a similar fashion as with the purified enzyme with some slight modifications. The reaction mixture consisted of: An estimated amount of 0.4 μM PTDH-mFMO in CFE, 1 mM Na-phosphite, 10 μL catalase (300 U), and 10 mM indole.

4.7. HPLC Analysis for Indigo Purity Measurements

Analyses were performed at 25 °C on an Alltima HP C18 column (3 μm, 4.6 by 150 mm I.D.; Grace, Reading, UK), equipped with a guard column. The mobile phase consisted of water containing 0.1% formic acid (eluent A) and acetonitrile (eluent B). The gradient program used was as follows:

10% B 2 min, 10% to 100% B (30 min), 100% B isocratic (5 min), and 100% to 10% B (3 min), followed by re-equilibration for 7 min. The flow rate was 0.5 mL/min and the injection volume were 20 μL. Commercially available indigo and indirubin were used as reference. Detector was set at 600 nm, since indigo has a maximal absorption at 610 nm whereas indirubin has a maximal absorption at 546 nm, as determined spectrophotometrically using standards.

5. Conclusions

Using rationally designed mutants we have shown it is possible to improve the flavin-containing monooxygenase mFMO on indole as substrate. A MISO library was used to expand the sequence space that was explored for improved activity, identifying further mutants with enhanced activity. We also showed that the wild-type enzyme already shows sufficient activity to perform conversions of indole to indigo. The set of improved mutants are good starting points for further engineering of the enzyme.

Supplementary Materials: Supplementary materials can be found at http://www.mdpi.com/1422-0067/20/24/6148/s1.

Author Contributions: Conceptualization, N.L. and M.W.F.; methodology, N.L. and H.L.v.B.; investigation, N.L. and H.L.v.B.; data curation, H.L.v.B.; writing—original draft preparation, N.L. and H.L.v.B.; writing—review and editing, all authors; funding acquisition, M.W.F.

Funding: This research was funded by NWO-LIFT project Indigreen.

Acknowledgments: We thank Arne Hundertmark and Tom Schepers for technical assistance and Hein Wijma for help in performing docking experiments.

Conflicts of Interest: The authors declare no conflict of interest.

Abbreviations

FRESCO	Framework for Rapid Enzyme Stabilization by Computational libraries
MISO	Multichange ISOthermal mutagenesis

References

1. Křížová, H. Natural dyes: Their past, present, future and sustainability. In *Recent Developments in Fibrous Material Science*; Křemenáková, D., Militký, J., Mishra, R., Eds.; Kanina: Prague, Czech Republic, 2015; pp. 59–71. ISBN 9788-0872-6945-9.
2. Steingruber, E. Indigo and Indigo Colorants. In *Ullmann's Encyclopedia of Industrial Chemistry*, 6th ed.; Wiley-VCH: Weinheim, Germany, 2004; Volume 19, p. 58. ISBN 9783-5273-0673-2.
3. Perkin, F. The Present Condition of the Indigo Industry. *Nature* **1900**, *63*, 7–9. [CrossRef]
4. Franssen, M.C.R.; Kircher, M.; Wohlgemuth, R. Industrial biotechnology in the chemical and pharmaceutical industry. In *Industrial Biotechnology. Sustainable Growth and Economic Success*; Soeteart, W., Vandamme, E.J., Eds.; Wiley-VCH: Weinheim, Germany, 2010; pp. 323–350. ISBN 978-3-527-31442-3.
5. Ensley, B.D.; Ratzkin, B.J.; Osslund, T.D.; Simon, M.J.; Wackett, L.P.; Gibson, D.T. Expression of naphthalene oxidation genes in *Escherichia coli* results in the biosynthesis of indigo. *Science* **1983**, *222*, 167–169. [CrossRef] [PubMed]
6. Kim, J.Y.; Lee, K.; Kim, Y.; Kim, C.K.; Lee, K. Production of dyestuffs from indole derivatives by naphthalene dioxygenase and toluene dioxygenase. *Lett. Appl. Microbiol.* **2003**, *36*, 343–348. [CrossRef] [PubMed]
7. Berry, A.; Dodge, T.C.; Pepsin, M.; Weyler, W. Application of metabolic engineering to improve both the production and use of biotech indigo. *J. Ind. Microbiol. Biotechnol.* **2002**, *28*, 127–133. [CrossRef]
8. Vaillancourt, F.H.; Labbe, G.; Drouin, N.M.; Fortin, P.D.; Eltis, L.D. The mechanism-based inactivation of 2,3-dihydroxybiphenyl 1,2-dioxygenase by catecholic substrates. *J. Biol. Chem.* **2002**, *277*, 2019–2027. [CrossRef]
9. Kim, J.; Lee, P.; Jung, E.; Kim, B. In vitro characterization of CYP102G4 from *Streptomyces cattleya*: A self-sufficient P450 naturally producing indigo. *Biochim. Biophys. Acta Proteins Proteomics* **2018**, *1866*, 60–67. [CrossRef]

10. Rosić, N.N. DNA shuffling of cytochromes P450 for indigoid pigment production. *Methods Mol. Biol.* **2013**, *987*, 205–224. [CrossRef]
11. Cheng, L.; Yin, S.; Chen, M.; Sun, B.; Hao, S.; Wang, C. Enhancing indigo production by over-expression of the styrene monooxygenase in *Pseudomonas putida*. *Curr. Microbiol.* **2016**, *73*, 248–254. [CrossRef]
12. Heine, T.; Großmann, C.; Hofmann, S.; Tischler, D. Indigoid dyes by group E monooxygenases: Mechanism and biocatalysis. *Biol. Chem.* **2019**, *400*, 939–950. [CrossRef]
13. Choi, H.S.; Kim, J.K.; Cho, E.H.; Kim, Y.C.; Kim, J.I.; Kim, S.W. A novel flavin-containing monooxygenase from *Methylophaga* sp strain SK1 and its indigo synthesis in *Escherichia coli*. *Biochem. Biophys. Res. Commun.* **2003**, *306*, 930–936. [CrossRef]
14. Han, G.H.; Shin, H.; Kim, S.W. Optimization of bio-indigo production by recombinant *E. coli* harboring fmo gene. *Enzyme Microb. Technol.* **2008**, *42*, 617–623. [CrossRef]
15. Alfieri, A.; Malito, E.; Orru, R.; Fraaije, M.W.; Mattevi, A. Revealing the moonlighting role of NADP in the structure of a flavin-containing monooxygenase. *Proc. Natl. Acad. Sci. USA* **2008**, *105*, 6572–6577. [CrossRef] [PubMed]
16. Furst, M.J.L.J.; Martin, C.; Loncar, N.; Fraaije, M.W. Experimental protocols for generating focused mutant libraries and screening for thermostable proteins. *Methods Enzymol.* **2018**, *608*, 151–187. [CrossRef] [PubMed]
17. Wijma, H.J.; Furst, M.J.L.J.; Janssen, D.B. A computational library design protocol for rapid improvement of protein stability: FRESCO. *Methods Mol. Biol.* **2018**, *1685*, 69–85. [CrossRef]
18. Mitchell, L.A.; Cai, Y.; Taylor, M.; Noronha, A.M.; Chuang, J.; Dai, L.; Boeke, J.D. Multichange isothermal mutagenesis: A new strategy for multiple site-directed mutations in plasmid DNA. *ACS Synth. Biol.* **2013**, *2*, 473–477. [CrossRef]
19. Furst, M.J.L.J.; Boonstra, M.; Bandstra, S.; Fraaije, M.W. Stabilization of cyclohexanone monooxygenase by computational and experimental library design. *Biotechnol. Bioeng.* **2019**, *116*, 2167–2177. [CrossRef]
20. Garcia-Macias, P.; John, P. Formation of natural indigo derived from woad (*Isatis tinctoria* L.) in relation to product purity. *J. Agric. Food Chem.* **2004**, *52*, 7891–7896. [CrossRef]
21. Cho, H.J.; Cho, H.Y.; Kim, K.J.; Kim, M.H.; Kim, S.W.; Kang, B.S. Structural and functional analysis of bacterial flavin-containing monooxygenase reveals its ping-pong-type reaction mechanism. *J. Struct. Biol.* **2011**, *175*, 39–48. [CrossRef]
22. Beaty, N.B.; Ballou, D.P. Transient kinetic study of liver microsomal FAD-containing monooxygenase. *J. Biol. Chem.* **1980**, *255*, 3817–3819.
23. Sheng, D.; Ballou, D.P.; Massey, V. Mechanistic studies of cyclohexanone monooxygenase: Chemical properties of intermediates involved in catalysis. *Biochemistry* **2001**, *40*, 11156–11167. [CrossRef]
24. Fordwour, O.B.; Luka, G.; Hoorfar, M.; Wolthers, K.R. Kinetic characterization of acetone monooxygenase from *Gordonia* sp. strain TY-5. *AMB Express* **2018**, *8*, 181. [CrossRef] [PubMed]
25. Orru, R.; Pazmiño, D.E.; Fraaije, M.W.; Mattevi, A. Joint functions of protein residues and NADP(H) in oxygen activation by flavin-containing monooxygenase. *J. Biol. Chem.* **2010**, *285*, 35021–35028. [CrossRef] [PubMed]
26. Li, C.Y.; Chen, X.L.; Zhang, D.; Wang, P.; Sheng, Q.; Peng, M.; Xie, B.B.; Qin, Q.L.; Li, P.Y.; Zhang, X.Y.; et al. Structural mechanism for bacterial oxidation of oceanic trimethylamine into trimethylamine N-oxide. *Mol. Microbiol.* **2017**, *103*, 992–1003. [CrossRef] [PubMed]
27. Rioz Martinez, A.; Kopacz, M.; de Gonzalo, G.; Torres Pazmiño, D.E.; Gotor, V.; Fraaije, M.W. Exploring the biocatalytic scope of a bacterial flavin-containing monooxygenase. *Org. Biomol. Chem.* **2011**, *9*, 1337–1341. [CrossRef]
28. Forneris, F.; Orru, R.; Bonivento, D.; Chiarelli, L.R.; Mattevi, A. *Thermo*FAD, a *Thermofluor*®-adapted flavin ad hoc detection system for protein folding and ligand binding. *FEBS J.* **2009**, *276*, 2833–2840. [CrossRef]

International Journal of
Molecular Sciences

MDPI

Article

Engineered Glucose Oxidase Capable of Quasi-Direct Electron Transfer after a Quick-and-Easy Modification with a Mediator

Nanami Suzuki [1,†], Jinhee Lee [2,†], Noya Loew [2], Yuka Takahashi-Inose [3], Junko Okuda-Shimazaki [2], Katsuhiro Kojima [3], Kazushige Mori [3], Wakako Tsugawa [1] and Koji Sode [2,*]

1 Department of Biotechnology and Life Science, Graduate School of Engineering, Tokyo University of Agriculture and Technology, 2-24-16 Naka-cho, Koganei, Tokyo 184-8588, Japan; nsuzuki@protonmail.com (N.S.); tsugawa@cc.tuat.ac.jp (W.T.)

2 Joint Department of Biomedical Engineering, The University of North Carolina at Chapel Hill and North Carolina State University, Chapel Hill, NC 27599, USA; jh.lee@unc.edu (J.L.); noya-loew@rs.tus.ac.jp (N.L.); jokudas@email.unc.edu (J.O.-S.)

3 Ultizyme International Ltd., 3-9-5. Taihei, Sumida, Tokyo 130-0012, Japan; yukappe1122@gmail.com (Y.T.-I.); katsuhiro.kojima@gmail.com (K.K.); mori_ka1213@yahoo.co.jp (K.M.)

* Correspondence: ksode@email.unc.edu; Tel.: +1-919-966-3550

† These authors contributed equally to this work.

Received: 3 January 2020; Accepted: 5 February 2020; Published: 8 February 2020

Abstract: Glucose oxidase (GOx) has been widely utilized for monitoring glycemic levels due to its availability, high activity, and specificity toward glucose. Among the three generations of electrochemical glucose sensor principles, direct electron transfer (DET)-based third-generation sensors are considered the ideal principle since the measurements can be carried out in the absence of a free redox mediator in the solution without the impact of oxygen and at a low enough potential for amperometric measurement to avoid the effect of electrochemically active interferences. However, natural GOx is not capable of DET. Therefore, a simple and rapid strategy to create DET-capable GOx is desired. In this study, we designed engineered GOx, which was made readily available for single-step modification with a redox mediator (phenazine ethosulfate, PES) on its surface via a lysine residue rationally introduced into the enzyme. Thus, PES-modified engineered GOx showed a quasi-DET response upon the addition of glucose. This strategy and the obtained results will contribute to the further development of quasi-DET GOx-based glucose monitoring dedicated to precise and accurate glycemic control for diabetic patient care.

Keywords: glucose oxidase; direct electron transfer; amine-reactive phenazine ethosulfate; glucose sensor; glycemic level monitoring

1. Introduction

Glucose oxidase (β-D-glucose: oxygen 1-oxidoreductases, E.C. 1.1.3.4, GOx) is a flavoprotein that catalyzes the oxidation of β-D-glucose at its hydroxyl group linked to the carbon one using molecular oxygen as the electron acceptor to produce D-glucono-1,5-lactone and hydrogen peroxide (Scheme 1). Since the first report by Clark and Lyons on an enzyme electrode employing GOx and an oxygen electrode for glucose monitoring [1], extensive studies have been carried out to develop improved enzyme-based systems for monitoring glycemic levels. The reason for the popularity of GOx lies in its commercial availability, high activity, and substrate specificity against glucose.

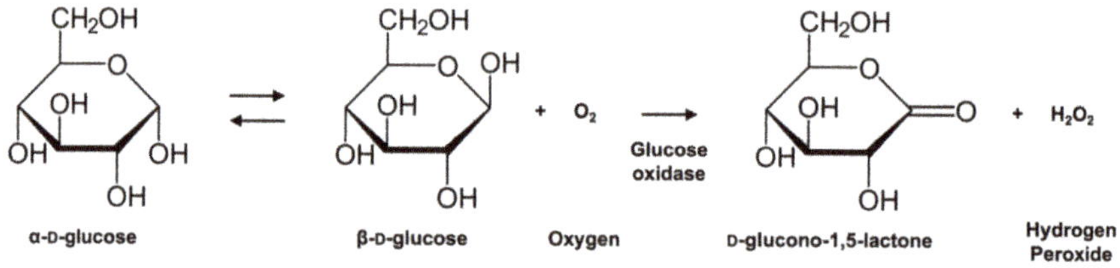

Scheme 1. Reaction scheme of glucose oxidation catalyzed by glucose oxidase.

Electrochemical glucose biosensors can be divided into three principles (Scheme 2). The "first-generation" glucose sensing principle utilizes molecular oxygen as the primary electron acceptor, and the glucose concentration is determined by the consumption of oxygen or generation of hydrogen peroxide. In the "second-generation" principle, artificial electron acceptors (also referred to as electron mediators) replace oxygen in the reaction [2], which enables avoiding the production of harmful hydrogen peroxide. Furthermore, due to the lower redox potential compared to that of hydrogen peroxide, the use of certain mediators enables us to decrease the applied potential for biosensors based on glucose oxidase. However, when oxidase is employed, this reaction is inherently affected by the oxygen concentration in a sample. Accordingly, glucose dehydrogenase (GDH) has become the major enzyme for glucose sensing in the second-generation principle. However, because of GDH's broad substrate specificity, which can lead to potentially fatal errors in glucose sensing [3], GOx has still been the focus due to its high specificity. Due to the advantages of GOx over GDH, studies are ongoing to engineer GOx into dehydrogenase [4–7]. In the "third-generation" sensor principle, glucose dehydrogenases capable of direct electron transfer (DET) are employed. Thereby, electrons derived from glucose oxidation are transferred directly to the electrode. Glucose sensor employing DET-type enzymes has several advantages comparing mediator type glucose sensors, e.g., a smaller number of essential elements for sensing, a smaller number of essential reactions for detection, operation at a lower applying potential than those utilizing hydrogen peroxide, and a major kind of mediator used in second-generation principle such as ferrocene, ferricyanide, and ruthenium complexes [8,9]. This low-applying potential contributes to eliminate interference by electrochemically active ingredients in the samples. Therefore, the DET-based enzyme sensor principle has been considered ideal for redox enzyme-based electrochemical sensors.

Considering the superior enzymatic property of the "gold standard," several efforts have been made to realize DET GOx-based glucose sensing. However, the 3D structure of GOx revealed that the redox cofactor, flavin adenine dinucleotide (FAD) is buried deeply within the protein. In addition, previous reports [10,11] attributed the voltammograms associated with GOx to free FAD, which suggests that only the free FAD of GOx showed voltammetric signals, whereas bound FAD did not, and bound FAD undergoes electron transfer at a more positive potential not observed in voltammetry [12]. Therefore, the only approach that realizes DET for GOx-based glucose sensing is the modification of the redox species onto the enzyme surface. Since the 1980s, GOx modifications by covalently attaching redox mediators have been reported to achieve quasi-DET using derivatives of ferrocene [13–21], phenothiazine [22], ruthenium [23], tetrathiafulvalene [24], and phenoxazine [25] to modify GOx. However, the modification procedure requires complex and laborious steps to introduce functional groups for mediator modification. In addition, no report has demonstrated the rational design of residues where redox mediators should be modified on GOx.

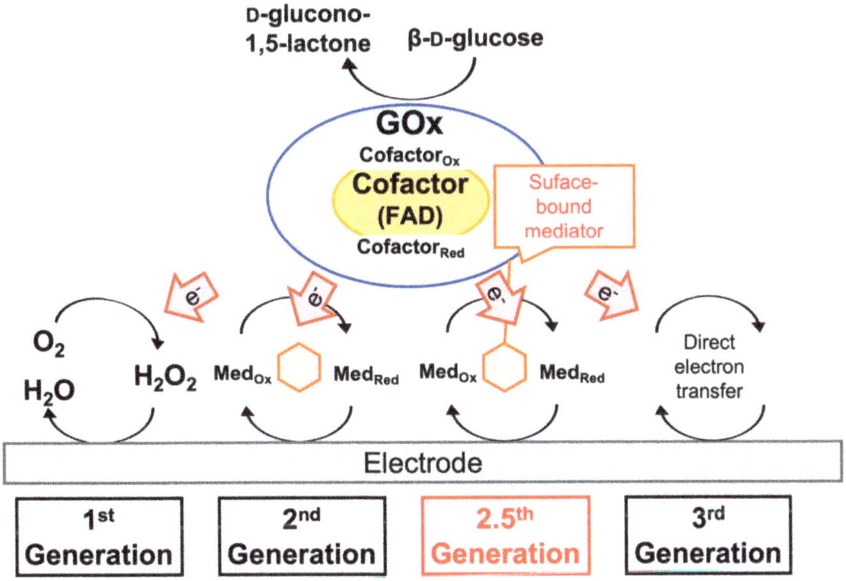

Scheme 2. Schematic illustration of the glucose sensor generations using glucose oxidase (GOx).

To achieve a versatile and simple method for enzyme modification, we previously reported a rapid and single-step conjugation using 1-[3-(succinimidyloxycarbonyl)propoxy]-5-ethylphenazinium trifluoromethanesulfonate (or amine-reactive phenazine ethosulfate; arPES) [26,27] onto the surface of the redox enzyme. The attachment of these redox mediators to redox enzymes enables intramolecular electron transfer from the enzyme redox center to the mediator and intermolecular electron transfer from the mediator to an electrode. This electron transfer via the mediator bound to the enzyme surface can be described as quasi-direct electron transfer (quasi-DET) and "2.5th generation" principle. arPES harbors two parts, i.e., a succinimide group and mediator, namely, methoxy PES, which is a stable, relatively low redox potential mediator. Upon mixing arPES and an enzyme, the primary amine groups on the enzyme attack the carboxylate ester carbon of the succinimide group of arPES and form amide bonds at room temperature.

Therefore, in this study, we aimed to design an engineered GOx suitable for quasi-DET-type sensor applications through the quick and easy modification of mediators using arPES. As the enzyme, GOx from *Aspergillus niger* was chosen, and amino acid substitution was designed based on the 3D structure to introduce a lysine residue in GOx, which was modified by arPES. After investigating the characteristics of the engineered GOx, PES-modified GOx was prepared, and a quasi-DET-type glucose sensor was constructed.

2. Results and Discussion

2.1. Engineering of GOx for PES Modification

2.1.1. Investigation of the Availability of Wild-Type *Aspergillus niger*-Derived GOx for Redox Mediator Modification

First, we investigated whether wild-type GOx from *A. niger* (*An*GOx-WT) is readily available for modification by PES to show a quasi-DET response. *An*GOx-WT was modified with PES, and the dye-mediated dehydrogenase activity was determined with an electron mediator, phenazine methosulfate (PMS), and a terminal electron acceptor, 3-(4,5-dimethyl-2-thiazolyl)-2,5-diphenyl-2*H*-tetrazolium bromide (MTT). When PMS and MTT were used together

(PMS/MTT system), the PMS functioned as the primary electron acceptor of the enzyme and mediated electrons to the bulky MTT. We previously reported that MTT alone did not function as the electron acceptor for a variety of redox enzymes. However, the presence of PMS facilitated the reduction of MTT by the oxidation substrate [26]. Considering that both PMS and MTT are negatively charged in their oxidative status, we assumed that the low reactivity of MTT as the primary electron acceptor might be attributed to its bulkiness when compared with PMS. Therefore, we have used the availability of MTT as the sole electron acceptor in the solution (MTT system) as the indication that PES was successively modified on the surface of redox enzymes, where internal electron transfer from a deeply buried cofactor to PES occurred. Consequently, bulky MTT can be reduced [26]. Since GOx was revealed that it did not utilize MTT as the primary electron acceptor, we followed our previous empirical experimental condition to analyze the modification of PES on the surface of GOx and evaluate the availability of the arPES modification procedure to prepare quasi-DET type GOx. We assumed that successive PES modification might enable PES-modified GOx to use MTT as the primary electron acceptor to show dye-mediated glucose dehydrogenase activity.

The results are shown in Table 1. The PES-modified *An*GOx-WT revealed almost identical dye-mediated dehydrogenase activity as that of unmodified and PES-modified *An*GOx-WT when the PMS/MTT system was employed, which indicates that the PES-modified GOx maintained its original activity. However, in the MTT system, only low dehydrogenase activity was observed after the modification of PES, which was indistinguishable from that of the intact *An*GOx-WT.

Table 1. Dehydrogenase activity of unmodified and phenazine ethosulfate (PES)-modified glucose oxidases (GOxs) (wild-type; WT and mutant; I489K) determined with the phenazine methosulfate (PMS)/3-(4,5-dimethyl-2-thiazolyl)-2,5-diphenyl-2*H*-tetrazolium bromide (MTT) system and MTT system.

Enzyme	PES Modification	Dehydrogenase Activity [U/mg]	
		PMS/MTT System	MTT System
*An*GOx-WT	−	16	n.d.*
	+	18	5.3×10^{-2}
*An*GOx-I489K	−	25	7.5×10^{-3}
	+	19	10

* Not detected.

To investigate the presence of the PES molecule on the surface of *An*GOx-WT after PES modification, we further characterized the enzyme on an electrode. The cyclic voltammograms of PES-modified *An*GOx-WT clearly showed oxidation and reduction peaks at approximately -100 mV, which revealed the presence of PES on the surface of the enzyme (Figure S1c), which were not observed from the electrode with PES-unmodified intact *An*GOx-WT (Figure S1a). These results indicate that PES was successfully modified on the surface of GOx and was electrochemically active. Considering that no catalytic current was observed in the presence of glucose (Figure S1d), together with the results of the previously mentioned dye-mediated dehydrogenase activity investigations using the MTT system, the position of PES modification on *An*GOx-WT was not suitable to accept electrons from its redox cofactor, FAD, in the catalytic center.

2.1.2. Identification of the Appropriate Site for Substitution with a Lysine Residue

To locate PES at a suitable position for the electron relay, we designed an additional lysine residue, where arPES should make a covalent bond. To determine the position, we compared the structure of *An*GOx (PDB ID: 1CF3) with a model structure of *Botryotinia fuckeliana*-derived GDH (*Bf*GDH), which has a structure similar to that of *An*GOx but shows a quasi-DET response after PES modification. A closer look at the position of lysine residues in the crystal structures of *An*GOx (1CF3), a structural model of *Bf*GDH, and the crystal structure of *A. flavus*-derived GDH (*Af*GDH, PDB ID: 4YNU), on which the model was based, shows that the residues are distributed very differently, despite the similar overall structure (Figure 1). However, there is one notable residue that is located at what appears to be the entrance of a pathway to the catalytic center. This residue is a lysine in both GDHs (K483 in *Bf*GDH, K477 in *Af*GDH) and an isoleucine in *An*GOx (I489). Furthermore, this residue appears to be located close enough to the FAD cofactor to receive electrons. Therefore, this residue appears to be an ideal candidate for attaching PES to function as an electron relay between the FAD and the terminal electron acceptor.

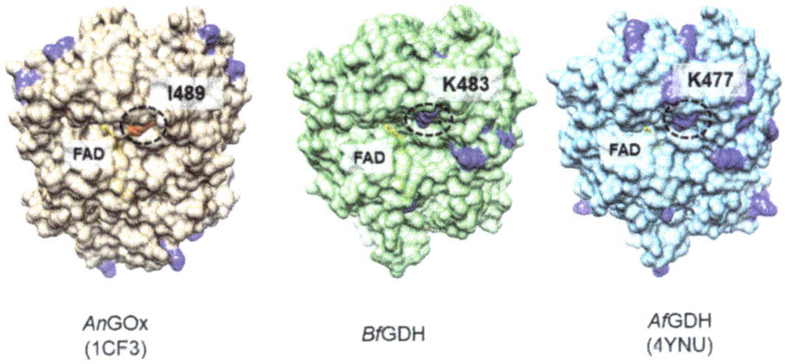

Figure 1. Comparison of the positions of lysine residues in *Aspergillus niger* derived glucose oxidase (*An*GOx) (PDB ID: 1CF3), *Botryotinia fuckeliana* derived glucose dehydrogenase (*Bf*GDH) (model), and *A.flavus* derived GDH (*Af*GDH) (PDB ID: 4YNU). Lysine residues are shown in dark blue. In *Bf*GDH and *Af*GDH, a lysine residue (K483, K477, circled) is located at the entrance of what appears to be a pathway to the active center. In *An*GOx, an isoleucine residue (I489, circled) is located at this position.

2.2. *Construction and Characterization of the GOx Mutant for Quasi-DET after PES Modification*

The isoleucine residue I489 in *An*GOx was substituted with a lysine residue in the *An*GOx-I489K mutant. Kinetic parameters of the mutant are shown in the Supplementary Material (Table S1). Notably, the modification of *An*GOx-I489K with PES significantly increased the activity value obtained with the MTT system (Table 1), which indicated the successful formation of an electron relay from FAD in the catalytic center to the enzyme surface. Additionally, a similar K_m value was observed with *An*GOx-I489K dehydrogenase activity before and after PES modification (Table S1). Therefore, it appears that the attachment of PES to the substrate entrance did not alter the accessibility of glucose to the active center.

Cyclic voltammograms of PES-modified *An*GOx-I489K clearly showed oxidation and reduction peaks at approximately −100 mV, which indicates the presence of PES on the enzyme surface (Figure 2c). In the presence of glucose, a catalytic current, i.e., an increase in the current at potentials higher than that of the oxidation peak of PES, was observed when we used PES-modified *An*GOx-I489K (Figure 2d), and no such catalytic current was observed with unmodified *An*GOx-I489K (Figure 2b). The observed reduction peak decrease at −0.4 V upon addition of glucose, which would be attributed to the decrease of local oxygen concentration caused by oxygen consumption by immobilized GOx. This result confirms that *An*GOx-I489K acquired quasi-DET ability after modification with PES. PES bound to the lysine residue at position 489 provided the missing link in the electron relay from FAD to the electrode.

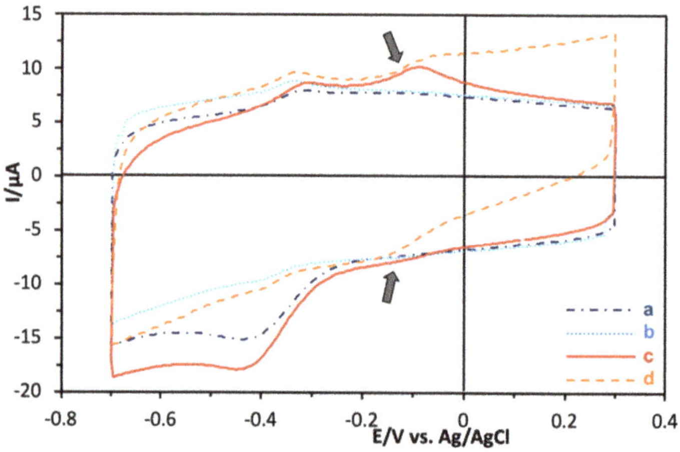

Figure 2. Cyclic voltammograms of (**a**) unmodified *An*GOx-I489K in the absence of glucose. (**b**) Unmodified *An*GOx-I489K in the presence of 33 mM glucose. (**c**) PES-modified *An*GOx-I489K in the absence of glucose. (**d**) PES-modified *An*GOx-I489K in the presence of 33 mM glucose. Scan rate 50 mV/s. Arrows: peaks due to PES.

2.3. Characterization of Electrodes with PES-Modified AnGOx-I489K

Lastly, we investigated the quasi-DET response against various glucose concentrations. Chronoamperometric measurements using the electrode with immobilized PES-modified *An*GOx-I489K showed clear and stable responses for the addition of glucose at an operating potential of 0 mV vs. Ag/AgCl (Figure 3A). A plot of the response current against the glucose concentration showed a linear increase until approximately 3 mM and saturation at glucose concentrations higher than approximately 17 mM (Figure 3B). The response on the electrode gave the apparent K_m value of 2.3 mM and I_{max} value of 1.4 µA even though the mutation nor PES modification did not change the K_m value of GOx (Table S1). This smaller apparent K_m value observed in the electrochemical measurement was assumed to be the difference of a rate limiting step. In the homogeneous solution, the reductive half reaction of the cofactor was the rate limiting step. However, in the electrochemical measurement, the electron transfer between quasi-DET GOx and the electrode might be limited. Essentially, the rate limiting step in the electrochemical reaction with PES modified GOx was assumed at the oxidative half reaction where reduced PES was oxidized by the electrode. Therefore, the apparent K_m value observed on the electrode did not match the actual K_m value determined with the redox dye.

The cyclic voltammograms (Figure 2) and chronoamperometric measurements (Figure 3A,B) confirm that PES-modified *An*GOx-I489K is capable of quasi-direct electron transfer due to PES modification of the introduced lysine residue. Based on these results, it seems that, after the FAD cofactor is reduced by glucose, it is re-oxidized by the PES bound to the lysine residue at position 489. The reduced PES is then oxidized directly at the electrode or possibly via other PES molecules bound at the surface of GOx.

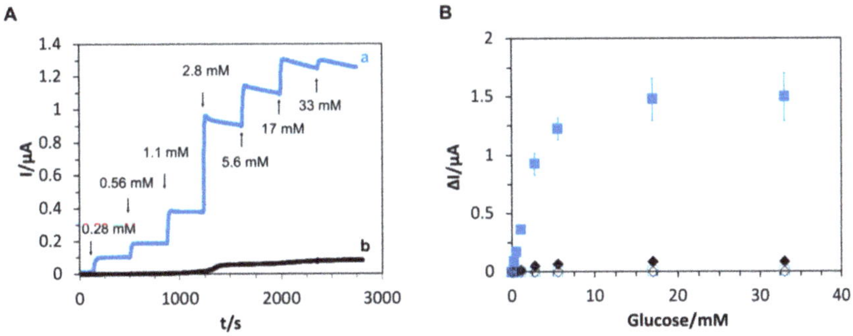

Figure 3. (**A**) Time courses of chronoamperometric measurements with (a) PES-modified *An*GOx-I489K and (b) PES-modified *An*GOx-WT. Application potential 0 mV vs. Ag/AgCl. Arrows indicate the addition of glucose. Concentrations over arrows are of the total glucose after addition. (**B**) Dependence of response current on glucose concentration for PES-modified *An*GOx-I489K (filled squares), PES-modified *An*GOx-WT (filled diamonds), unmodified *An*GOx-I489K (open squares), and unmodified *An*GOx-WT (open diamonds). Application potential 0 mV vs. Ag/AgCl.

3. Materials and Methods

3.1. Materials

arPES was kindly donated by Dojindo Laboratories Co. Ltd. (Kumamoto, Japan). Phenazine methosulfate (PMS), D(+)-glucose, potassium dihydrogen phosphate, dipotassium hydrogen phosphate and Triton X-100 were purchased from Kanto Chemical Co. Inc. (Tokyo, Japan). 3-(4,5-Dimethyl-2-thiazolyl)-2,5-diphenyl-2*H*-tetrazolium bromide (MTT) was purchased from Dojindo Laboratories Co. Ltd. D-(+)-trehalose dihydrate and glutaraldehyde were purchased from FUJIFILM Wako Pure Chemical Corporation (Osaka, Japan). *N*-[Tris(hydroxymethyl)methyl]glycine (tricine) was purchased from Sigma-Aldrich Co. LLC (St. Louis, MO, USA). Ketjen black (ECP600JD), was obtained from Mitsubishi Chemical Corporation (Tokyo, Japan).

3.2. Identification of the Appropriate Mutation Site by Comparing the GOx and GDH Structures

A 3D structural model of *Botryotinia fuckeliana*-derived GDH (*Bf*GDH) was generated based on the crystal structure of FAD-dependent glucose dehydrogenase derived from *A. flavus* (*Af*GDH, PDB ID: 4YNU) [28]. The amino acid sequences of GOx derived from *A. niger* (*An*GOx, PDB ID: 1CF3) and of the GDHs were aligned using the multiple sequence alignment software ClustalW (http://www.clustal.org) [29]. By comparing the structures of GOx and the GDHs, lysine residues in the GDHs were identified, which are close to what appears to be the entrance of pathways leading to the catalytic center and which have corresponding non-lysine residues in the GOx.

3.3. Expression Vector Preparation and Recombinant Expression of GOxs

The structural gene of wild-type GOx was prepared as previously described [7]. The gene was inserted into the multiple cloning site of the expression vector pET30c(+) (Merck KGaA, Darmstadt, Germany). Expression vectors of GOx mutants were prepared by site-directed mutagenesis using the QuikChange Mutagenesis Kit (Agilent Technology Inc., Santa Clara, CA, USA). The correctness of the mutations was confirmed using the ABI Prism 3100 Genetic Analyzer (Applied Biosystems, Foster City, CA, USA). Expression and refolding of GOxs (both wild type and mutant) were carried out as previously described [30].

3.4. PES Modification of Enzymes

GOxs (wild-type and mutant) were modified using arPES, as previously described [26] with some changes. GOx and arPES were mixed in 20 mM tricine buffer (pH 8.3) to 13 μM (based on catalytic centers) and 2.0 mM, respectively, and incubated at 25 °C for 2 h while shaking at 1200 rpm. Afterward, the sample was ultra-filtered to remove excess arPES and exchange the buffer to 20 mM potassium phosphate buffer (P.P.B.) using Amicon Ultra-0.5 30K centrifugal filters (molecular cut-off, 30 kDa) (Sigma-Aldrich Co. LLC, St. Louis, MO, USA).

3.5. Enzyme Activity

Two types of dehydrogenase activity assays were performed, which were named the PMS/MTT system and MTT system. For the PMS/MTT system, the increase in absorbance at 565 nm (from the formation of formazan dye due to the reduction of MTT) of a mixture of GOx, 100 mM glucose, 0.6 mM PMS, and 1 mM MTT in 20 mM P.P.B. was monitored. For the MTT system, the increase in absorbance at 565 nm of a mixture of GOx, 100 mM glucose, and 1 mM MTT in 20 mM P.P.B. was monitored. The reduction of 1 μmol MTT in 1 min, corresponding to the oxidation of 1 μmol/min glucose, was defined as 1 U dehydrogenase activity.

3.6. Preparation of Enzyme Electrodes

First, 0.54 μL of enzyme ink containing 1.5 mg/mL GOx (unmodified or PES-modified), 0.5% trehalose, 0.6% Ketjen black, and 1.2% Triton X-100 in 20 mM P.P.B. was dropped on a glassy carbon electrode (3.0 mm diameter, BAS Inc., Tokyo, Japan), and dried at 25 °C. Next, the enzyme was cross-linked in glutaraldehyde vapor for 1 h at room temperature. The electrodes were stored at 25 °C until use.

3.7. Electrochemical Evaluation

Before use, the electrodes were equilibrated in the electrolyte (100 mM P.P.B., pH 7.0) for at least 20 min. A platinum wire (TANAKA Kikinzoku K.K., Tokyo, Japan) was used as a counter electrode and Ag/AgCl/3 M NaCl as a reference electrode (BAS Inc., Tokyo, Japan). An SP-150 potentiostat (Bio-Logic Science Instruments, Seyssinet-Pariset, France) was used for electrochemical measurements. Cyclic voltammetry was performed for a potential range of −0.7 V to 0.3 V at a scan rate of 50 mV/s. For the chronoamperometric measurements, a potential of 0 V was applied. Glucose was added to the final concentrations of 0, 0.28, 0.56, 1.1, 2.8, 5.6, 17, and 33 mM.

4. Conclusions

By strategically designing the position of a lysine residue for PES modification by the structural comparison between GOx and FADGDHs, an engineered *An*GOx-I489K was constructed. *An*GOx-I489K was readily modified by arPES, which enables the enzyme to exhibit MTT system-dependent dye-mediated glucose dehydrogenase activity and an electrochemical response on the electrode upon the addition of glucose. These properties were distinct from those of PES-modified *An*GOx-WT. The enzyme electrode with PES-modified *An*GOx-I489K showed a response at a potential as low as 0 mV vs. Ag/AgCl, which enabled the elimination of an effect from electrochemically-active interference. Additionally, since a direct electron transfer-type sensor does not require a free electron mediator, we are able to simplify the architecture of a sensor.

In conclusion, the engineered GOx we show in this case has the advantage of quasi-DET-ability in addition to its original functionality (high specificity and activity). This achievement realized a revived version of the "gold standard" glucose sensing enzyme, GOx, which has been used in the first-generation and second-generation principles based on glucose sensors, which will be available for the "2.5th generation" based glucose sensors. The addition of quasi-DET ability to GOx make this enzyme the invaluable material to construct a mediator-free type glucose sensor strip, and an ingredient impact-free CGM sensor.

Supplementary Materials: Supplementary Materials can be found at http://www.mdpi.com/1422-0067/21/3/1137/s1.

Author Contributions: Conceptualization, W.T. and K.S. Experimental design, Y.T-I.., J.O.-S., and K.M. Data acquisition, N.S., Y.T.-I., J.O., K.K., and K.M. Data analysis and interpretation, N.S., J.L., Y.T.-I., K.K., K.M., W.T., and K.S. Writing—original draft preparation, J.L. and N.L. Writing—review and editing, K.S. All authors have read and agreed to the published version of the manuscript.

Funding: This research received no external funding.

Acknowledgments: The authors appreciate Dojindo Laboratories Co. Ltd. (Kumamoto, Japan) very much for their kind supply of arPES.

Conflicts of Interest: The authors declare no conflict of interest.

Abbreviations

FAD	Flavin adenine dinucleotide
GOx	Glucose oxidase
GDH	Glucose dehydrogenase
DET	Direct electron transfer
MTT	3-(4,5-Dimethyl-2-thiazolyl)-2,5-diphenyl-2*H*-tetrazolium bromide
PES	Phenazine ethosulfate
PMS	Phenazine methosulfate

References

1. Clark, L.C.; Lyons, C. Electrode systems for continuous monitoring in cardiovascular surgery. *Ann. N. Y. Acad. Sci.* **1962**, *102*, 29–45. [CrossRef] [PubMed]
2. Cass, A.E.G.; Davis, G.; Francis, G.D.; Hill, H.A.O.; Aston, W.J.; Higgins, I.J.; Plotkin, E.V.; Scott, L.D.L.; Turner, A.P.F. Ferrocene-Mediated enzyme electrode for amperometric determination of glucose. *Anal. Chem.* **1984**, *56*, 667–671. [CrossRef]
3. Ferri, S.; Kojima, K.; Sode, K. Review of glucose oxidases and glucose dehydrogenases: A bird's eye view of glucose sensing enzymes. *J. Diabetes Sci. Technol.* **2011**, *5*, 1068–1076. [CrossRef] [PubMed]
4. Tremey, E.; Stines-Chaumeil, C.; Gounel, S.; Mano, N. Designing an O_2-insensitive glucose oxidase for improved electrochemical applications. *ChemElectroChem* **2017**, *4*, 2520–2526. [CrossRef]
5. Horaguchi, Y.; Saito, S.; Kojima, K.; Tsugawa, W.; Ferri, S.; Sode, K. Engineering glucose oxidase to minimize the influence of oxygen on sensor response. *Electrochim. Acta* **2014**, *126*, 158–161. [CrossRef]
6. Gutierrez, E.A.; Mundhada, H.; Meier, T.; Duefel, H.; Bocola, M.; Schwaneberg, U. Reengineered glucose oxidase for amperometric glucose determination in diabetes analytics. *Biosens. Bioelectron.* **2013**, *50*, 84–90. [CrossRef]
7. Horaguchi, Y.; Saito, S.; Kojima, K.; Tsugawa, W.; Ferri, S.; Sode, K. Construction of mutant glucose oxidases with increased dye-mediated dehydrogenase activity. *Int. J. Mol. Sci.* **2012**, *13*, 14149–14157. [CrossRef]
8. Chaubey, A.; Malhotra, B.D. Mediated biosensors. *Biosens. Bioelectron.* **2002**, *17*, 441–456. [CrossRef]
9. Lee, I.; Loew, N.; Tsugawa, W.; Lin, C.; Probst, D.; La Belle, J.T.; Sode, K. The electrochemical behavior of a FAD dependent glucosedehydrogenase with direct electron transfer subunit by immobilizationon self-assembled monolayers. *Bioelectrochemistry* **2018**, *121*, 1–6. [CrossRef]
10. Bartlett, P.N.; Al-Lolage, F.A. There is no evidence to support literature claims of direct electron transfer (DET) for native glucose oxidase (GOx) at carbon nanotubes or graphene. *J. Electroanal. Chem.* **2018**, *819*, 26–37. [CrossRef]
11. Wilson, G.S. Native glucose oxidase does not undergo direct electron transfer. *Biosens. Bioelectron.* **2016**, *82*, Vii–Viii. [CrossRef] [PubMed]
12. Vogt, S.; Schneider, M.; Schafer-Eberwein, H.; Noll, G. Determination of the pH dependent redox potential of glucose oxidase by spectroelectrochemistry. *Anal. Chem.* **2014**, *86*, 7530–7535. [CrossRef]
13. Sampath, S.; Lev, O. Renewable, reagentless glucose sensor based on a redox modified enzyme and carbon-silica composite. *Electroanalysis* **1996**, *8*, 1112–1116. [CrossRef]

14. Schuhmann, W. Electron-transfer pathways in amperometric biosensors. Ferrocene-modified enzymes entrapped in conducting-polymer layers. *Biosens. Bioelectron.* **1995**, *10*, 181–193. [CrossRef]
15. Badia, A.; Carlini, R.; Fernandez, A.; Battaglini, F.; Mikkelsen, S.R.; English, A.M. Intramolecular electron-transfer rates in ferrocene-derivatized glucose-oxidase. *J. Am. Chem. Soc.* **1993**, *115*, 7053–7060. [CrossRef]
16. Ryabov, A.D.; Trushkin, A.M.; Baksheeva, L.I.; Gorbatova, R.K.; Kubrakova, I.V.; Mozhaev, V.V.; Gnedenko, B.B.; Levashov, A.V. Chemical attachment of organometallics to proteins in reverse micelles. *Angew. Chem. Int. Ed.* **1992**, *31*, 789–791. [CrossRef]
17. Schuhmann, W.; Ohara, T.J.; Schmidt, H.L.; Heller, A. Electron-transfer between glucose-oxidase and electrodes via redox mediators bound with flexible chains to the enzyme surface. *J. Am. Chem. Soc.* **1991**, *113*, 1394–1397. [CrossRef]
18. Bartlett, P.N.; Bradford, V.Q.; Whitaker, R.G. Enzyme electrode studies of glucose-oxidase modified with a redox mediator. *Talanta* **1991**, *38*, 57–63. [CrossRef]
19. Degani, Y.; Heller, A. Direct electrical communication between dhemically modified enzymes and metal-electrodes. 2. Methods for bonding electron-transfer relays to glucose-oxidase and D-amino-acid oxidase. *J. Am. Chem. Soc.* **1988**, *110*, 2615–2620. [CrossRef]
20. Degani, Y.; Heller, A. Direct electrical communication between chemically modified enzymes and metal-electrodes. 1. Electron-transfer from glucose-oxidase to metal-electrodes via electron relays, bound covalently to the enzyme. *J. Phys. Chem.* **1987**, *91*, 1285–1289. [CrossRef]
21. Bartlett, P.N.; Whitaker, R.G.; Green, M.J.; Frew, J. Covalent binding of electron relays to glucose-oxidase. *J. Chem. Soc. Chem. Comm.* **1987**, 1603–1604. [CrossRef]
22. Ban, K.; Ueki, T.; Tamada, Y.; Saito, T.; Imabayashi, S.; Watanabe, M. Fast electron transfer between glucose oxidase and electrodes via phenothiazine mediators with poly(ethylene oxide) spacers attached to the enzyme surface. *Electrochem. Commun.* **2001**, *3*, 649–653. [CrossRef]
23. Ryabova, E.S.; Goral, V.N.; Csoregi, E.; Mattiasson, B.; Ryabov, A.D. Coordinative approach to mediated electron transfer: Ruthenium complexed to native glucose oxidase. *Angew. Chem. Int. Ed.* **1999**, *38*, 804–807. [CrossRef]
24. Bartlett, P.N.; Booth, S.; Caruana, D.J.; Kilburn, J.D.; Santamaria, C. Modification of glucose oxidase by the covalent attachment of a tetrathiafulvalene derivative. *Anal. Chem.* **1997**, *69*, 734–742. [CrossRef]
25. Krikstopaitis, K.; Kulys, J.; Tetianec, L. Bioelectrocatalytical glucose oxidation with phenoxazine modified glucose oxidase. *Electrochem. Commun.* **2004**, *6*, 331–336. [CrossRef]
26. Hatada, M.; Loew, N.; Inose-Takahashi, Y.; Okuda-Shimazaki, J.; Tsugawa, W.; Mulchandani, A.; Sode, K. Development of a glucose sensor employing quick and easy modification method with mediator for altering electron acceptor preference. *Bioelectrochemistry* **2018**, *121*, 185–190. [CrossRef]
27. Hiraka, K.; Kojima, K.; Tsugawa, W.; Asano, R.; Ikebukuro, K.; Sode, K. Rational engineering of Aerococcus viridans L-lactate oxidase for the mediator modification to achieve quasi-direct electron transfer type lactate sensor. *Biosens. Bioelectron.* **2019**, *151*, 111974. [CrossRef]
28. Yoshida, H.; Sakai, G.; Mori, K.; Kojima, K.; Kamitori, S.; Sode, K. Structural analysis of fungus-derived FAD glucose dehydrogenase. *Sci. Rep.* **2015**, *5*, 13498. [CrossRef]
29. Thompson, J.D.; Higgins, D.G.; Gibson, T.J. CLUSTAL W: Improving the sensitivity of progressive multiple sequence alignment through sequence weighting, position-specific gap penalties and weight matrix choice. *Nucleic Acids Res.* **1994**, *22*, 4673–4680. [CrossRef]
30. Okurita, M.; Suzuki, N.; Loew, N.; Yoshida, H.; Tsugawa, W.; Mori, K.; Kojima, K.; Klonoff, D.C.; Sode, K. Engineered fungus derived FAD-dependent glucose dehydrogenase with acquired ability to utilize hexaammineruthenium(III) as an electron acceptor. *Bioelectrochemistry* **2018**, *123*, 62–69. [CrossRef]

International Journal of
Molecular Sciences

MDPI

Article

Manganese(II) Oxidizing Bacteria as Whole-Cell Catalyst for β-Keto Ester Oxidation

Juan Guo [1], Huan Guo [1], Jin Liu [1,*], Fangrui Zhong [1] and Yuzhou Wu [1,2,*]

[1] Hubei Key Laboratory of Bioinorganic Chemistry and Materia Medica, School of Chemistry and Chemical Engineering, Huazhong University of Science and Technology, Wuhan 430074, China; guojuan0201@163.com (J.G.); d201980126@hust.edu.cn (H.G.); chemzfr@hust.edu.cn (F.Z.)

[2] Max Planck Institute for Polymer Research, Ackermannweg 10, 55128 Mainz, Germany

* Correspondence: liujinlj1987@163.com (J.L.); wuyuzhou@hust.edu.cn (Y.W.)

Received: 14 February 2020; Accepted: 24 February 2020; Published: 2 March 2020

Abstract: Manganese oxidizing bacteria can produce biogenic manganese oxides (BMO) on their cell surface and have been applied in the fields of agriculture, bioremediation, and drinking water treatment to remove toxic contaminants based on their remarkable chemical reactivity. Herein, we report for the first time the synthetic application of the manganese oxidizing bacteria, *Pseudomonas putida* MnB1 as a whole-cell biocatalyst for the effective oxidation of β-keto ester with excellent yield. Differing from known chemical protocols toward this transformation that generally necessitate the use of organic solvents, stoichiometric oxygenating agents and complex chemical catalysts, our strategy can accomplish it simply under aqueous and mild conditions with higher efficiency than that provided by chemical manganese oxides. Moreover, the live MnB1 bacteria are capable of continuous catalysis for this C-O bond forming reaction for several cycles and remain proliferating, highlighting the favorable merits of this novel protocol for sustainable chemistry and green synthesis.

Keywords: *Pseudomonas putida* MnB1; biogenic manganese oxides; abiotic manganese oxides; α-Hydroxy-β-keto esters; whole-cell biocatalysis

1. Introduction

Developing sustainable biocatalytic processes for chemical synthesis has attracted considerable attention due to the ever-increasing environment concerns [1–3]. Conventional chemical production provides organic compounds that fulfil fundamental demands of modern society in pharmaceutical, agricultural, material and other fields, however, often at the expense of environment pollution and energy consumption. As such, biocatalysis provides a more favorable alternative considering its merits such as high catalytic activity and selectivity, mild reaction conditions (physiological pH and temperature), and environmental credentials (enzymes, organic solvent-free medium) [4–6]. In particular, whole-cell biocatalysis possesses unique advantages and extraordinary attractiveness. First, enzymes inside cells are to some extent in a protected environment and therefore often more stable than their isolated counterparts [7]. Besides, whole-cell biocatalysis integrates the benefits of enzyme cascades in a bacterial system and the fast proliferation of a living microbe, thus being more energy efficient, sustainable and easily recyclable [8]. However, the whole-cell catalytic reactions necessitate fast transportation of non-toxic substrates across the cell envelope to contact the enzymes, which essentially limits the substrate scope and reaction rate [9]. Therefore, novel strategies to utilize microorganisms for useful organic transformations are demanded to broaden the application of whole-cell biocatalysis in sustainable synthesis of fine chemicals.

Manganese dioxide (MnO_2) is a classic oxidant in organic synthesis with broad substrate scope and high reaction selectivity, as seen in alcohol oxidation, aromatization, oxidative coupling, and thiol oxidation [10–14]. In nature, biogenic manganese oxides (BMO) produced by Mn(II) oxidizing bacteria is widely present in soil and sediment, which has been extensively studied as a chemical catalyst or oxidizing reagent to remove various organic pollutants [15–17]. Of note, the main content of BMO is MnO_2, which was found to have even larger specific surface area and higher reactivity than chemically prepared equivalents [18,19]. BMO producing bacteria can be directly applied in the fields of agriculture, bioremediation, and drinking water treatment to remove toxic contaminants [20–23], exhibiting extraordinary advantages such as high efficiency, low cost and environmental safety. Moreover, since the BMO is produced on the surface of bacteria and even secretes to the environment, these microbes can catalyze reactions without requiring the cell uptake of substrates and thus might benefit the reaction kinetics. Despite remarkable advances in various fields, the use of Mn(II) oxidizing bacteria as a whole-cell catalyst for synthesizing fine chemicals has not been explored (Figure 1).

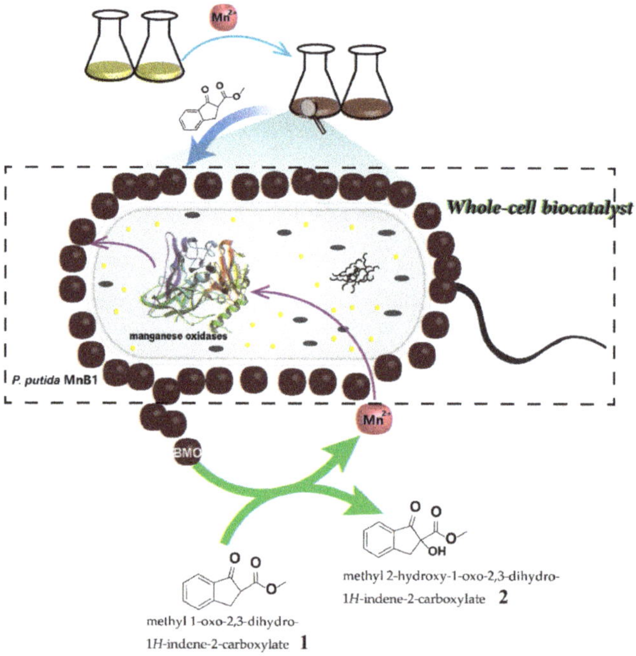

Figure 1. α-hydroxy-β-keto ester (**1**) by whole-cell biocatalysis based on biogenic manganese oxides (BMO).

Pseudomonas putida MnB1, one of the most studied Mn(II) oxidizing bacteria, is ubiquitous in freshwater and soil, and can be cultivated even in complicated environments [20]. It would oxidize Mn(II) in liquid and solid media to Mn(IV) and accumulate BMO precipitates on the cell surface [24]. The robustness of *P. putida* MnB1 lays the groundwork for their prospective synthetic application as potential biocatalyst. To prove the concept of Mn(II) oxidizing bacteria whole-cell biocatalysis for organic synthesis, α-hydroxylation of β-keto ester (**1**) (methyl 1-oxo-2,3-dihydro-1H-indene-2-carboxylate) was selected as model reaction. This reaction provides the most straightforward access to the α-hydroxy-β-dicarbonyl, an intriguing moiety commonly found in various biologically active natural products, agrochemicals, and pharmaceuticals [25–27]. Notably, a number of chemical protocols are available to accomplish this oxidation to yield product **2** [28–33]. For instance, Lu et al. reported a Brønsted acid catalytic method with nitrosobenzene as the oxygen source [28], and Meng and co-workers documented a Zr(IV)/organic peroxide system [30]. In general, the use of organic solvents and stoichiometric oxygenating agents were necessitated in conjunction with complex chemical catalysts, thus strongly compromising reaction economy and environmental friendliness. Herein, the BMO-based MnB1 catalyzed α-hydroxylation of β-keto ester (**1**) can be successfully achieved in water with superior performance than that of chemically produced MnO_2. Moreover, the live MnB1 bacteria can be recycled with ease and remain proliferating, thus they are capable of continuously catalyzing the conversion of substrates. Therefore, this is a sustainable whole-cell biocatalytic system for efficient oxidation reaction that holds marked advantages for industrial applications due to the high efficiency, low cost and potential for flow chemistry.

2. Results

2.1. Biogenic Manganese Oxides (BMO) Formation

The reaction was firstly tested with lyophilized powder of the BMO mineralized MnB1 bacteria (noted as dry m-MnB1). To prepare it, MnB1 was cultured in Lept medium containing 1 mM Mn^{2+} for mineralization and the formation of BMO was observed by the production of dark brown sediments after 35 h. The bacteria were continuously cultured for 5 d before harvesting by centrifugation and lyophilization. The Mn content in the dry m-MnB1 was determined to be 28.8% according to inductively coupled plasma optical emission spectrometer (ICP-OES) (Figure S1). This result is consistent with the respective literature data (18–30%) [18,19].

2.2. Reactivity Comparison of the Dry m-MnB1 with Chemical Manganese Oxides (CMO) toward the Oxidation of β-Keto Ester

2.2.1. Solvent Effects on Oxidation Rate

The dry m-MnB1 containing the BMO lyophilized together with MnB1 bacteria was mixed with β-keto ester (**1**) to test the oxidation reactivity. The performance of dry m-MnB1 was first compared with commercially available MnO_2 powder (CMO) in different media consisting of H_2O and acetonitrile (MeCN) under otherwise identical conditions. The Mn content was adjusted to be the same for both dry m-MnB1 and CMO. The oxidized product was characterized by ^1H NMR spectroscopy (Figure S2). The yield of product **2** was determined by high performance liquid chromatography (HPLC) using comparably normalized standards (Figure S3). It was found that the mixed solvent system comprised of 90% H_2O and 10% MeCN gave the highest yield of product 2 for both BMO and CMO (3 h, 94.1% and 54.7%, respectively) (Figure 2a). Increasing the organic proportion was detrimental. For instance, the respective yield dropped to 21.8% and 12.7% in MeCN/H_2O (3:1). However, a small amount of MeCN was still necessary to dissolve the organic substrate. Nevertheless, the aqueous system adopted herein is much more favorable than known protocols exclusively using organic solvents.

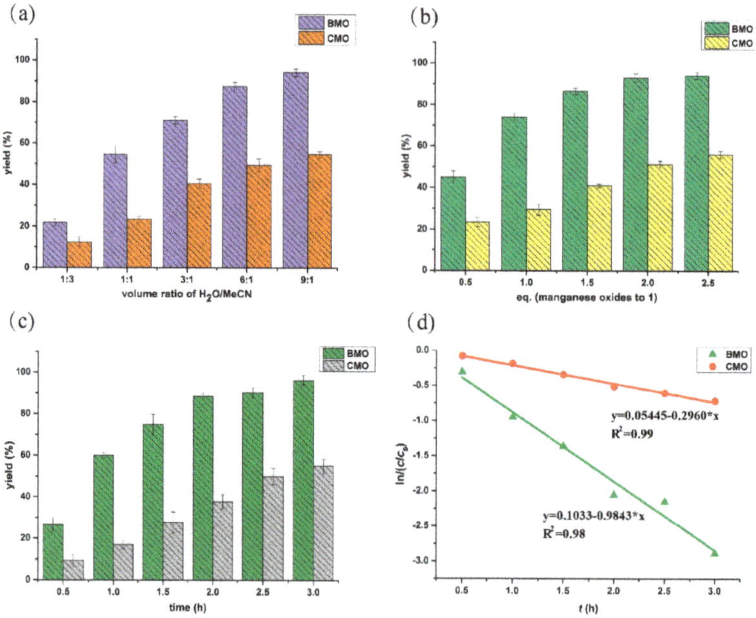

Figure 2. Comparing the synthesis of α-hydroxy-β-keto ester by the BMO (biogenic manganese oxides) from dry m-MnB1 and CMO (chemical manganese oxides): (**a**) Effects of reaction media consisting of H_2O/MeCN with different ratios; (**b**) Effects of dosage of BMO and CMO in H_2O/MeCN (9:1); (**c**) Correlation of yield of **2** with time; (**d**) Pseudo-first-order plots of α-hydroxylation of β-keto Ester by BMO and CMO in the first 3 h.

Notably, in all tested solvent conditions, BMO in dry m-MnB1 revealed significantly higher reactivity than CMO, and almost 2-fold higher yields were attainable in any given medium (Figure 2a). These results proved that the dry m-MnB1 possesses superior reactivity and selectivity than conventional CMO.

2.2.2. Influence of Dosage of Manganese Oxides on Reaction Rate

We subsequently explored the effects of dosage of manganese oxides on the oxidation of compound **1**. A mixture of H_2O/MeCN (9:1) was employed as the reaction medium and the reaction time was set at 1 h. As shown in Figure 2b, the ratio of manganese oxides to substrate **1** was found to significantly impact the oxidation process. As the dosage of BMO from dry m-MnB1 was increased from 0.5 eq. to 1.0 eq., the yield of **2** raised from 42% to 72%. The outcome could be further improved to 91% with 2.0 eq of oxidant, and it became marginally higher with further addition of BMO. These results suggest 2.0 eq. of BMO is adequate. As the substrate **1** dissolved in the reaction medium was originally 5.0 mM, the factor limiting it from approaching full conversion might be the extremely low concentration at this point. Analogously, the same set of experiments were conducted with CMO. Although the conversion of substrate **1** was also evidently promoted along the addition of oxidant, the yield of **2** was constantly lower than BMO under identical conditions, for instance, as seen from 56% yield versus 93% yield with 2.5 eq. of manganese oxides. The results also back the notion that the BMO from dry m-MnB1 is more efficient than the CMO for this oxidation reaction.

2.2.3. Reaction Kinetics

The kinetic parameters for both BMO and CMO mediated oxidation were studied (Figure 2c,d). It was found that α-hydroxylation of substrate **1** with both BMO from dry m-MnB1 and CMO in the first 3 h followed Pseudo-first order kinetics, which were expressed as in equation (1), with C: concentration of 1 (mmol/L) at time t (h), C_0: the initial concentration of **1** (mmol/L), and k: the rate constant (h^{-1}). By transforming Equation (1) into Equation (2),

$$C = C_0 \cdot e^{-k \cdot t} \quad (1)$$

$$ln\left(\frac{c}{c_0}\right) = -k \cdot t \quad (2)$$

the rate constant can be calculated from the slope of the graph as depicted in Figure 2d. An average value for the rate constant k of 0.9843 ± 0.0215 h^{-1} was calculated with BMO from dry m-MnB1, while on the contrary, the rate constant of CMO was merely 0.296 ± 0.0286 h^{-1} (Table 1). It is clear that the oxidation activity of the BMO from dry m-MnB1 is higher than CMO.

Table 1. Kinetic parameters of manganese oxides.

Data	Dried BMO [1]	CMO [2]
Rate constant (k, h^{-1})	0.9843 ± 0.0215	0.296 ± 0.0286
R^2	0.97~0.98	0.98~0.99

[1] BMO: biogenic manganese oxides; [2] CMO: chemical manganese oxides.

2.3. Effects of β-Keto Ester 1 on Bacteria Growth

In order to achieve the α-hydroxylation of **1** by directly using live MnB1 bacteria, we moved on to investigate the influence substrate **1** on the growth of *P. putida* MnB1 at different concentrations. As illustrated in Figure 3a and Figure S4, the MnB1 growth profile did not deviate notably from the standard strain incubation in the presence of 5.0 mM of substrate, while larger concentration (10.0 mM) exhibited obvious toxicity. Further increase of concentration to 15.0 and 20.0 mM both significantly suppressed the bacterial growth. Nevertheless, the bacteria were still proliferating even with 20.0 mM substrate, highlighting the robustness of MnB1.

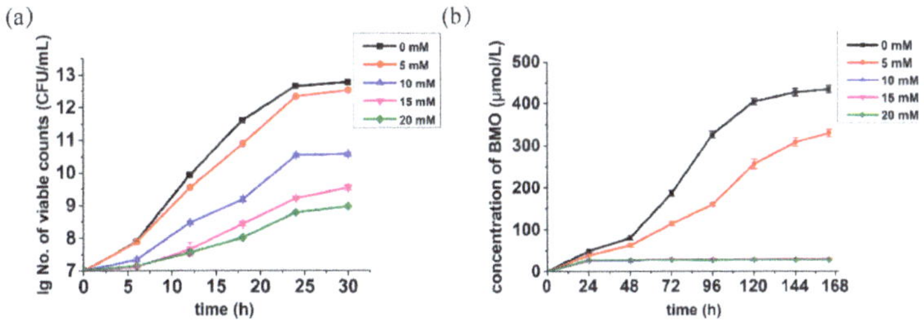

Figure 3. Effects of β-keto ester on the bacteria growth (**a**) measured by spread plate method and (**b**) the manganese mineralization; various colors refer to different concentrations of **1**; CFU: colony forming units.

2.4. Effects of β-Keto Ester 1 on Manganese Mineralization

The effects of various concentrations of **1** on the manganese mineralization by the model strain were also probed. However, the manganese mineralization of MnB1 was found to be much more sensitive to substrate **1**. A concentration of 5.0 mM has already shown a significant inhibition on mineralization process, which was found to be completely repressed at 10.0 mM or higher (Figure 3b). Since the generation of BMO by *P. putida* MnB1 was fundamentally accomplished via multicopper oxidase-catalyzed oxidation of Mn(II) into Mn(VI) in cell extracts within its active center with three Cu(II) sites [34–36], and the β-keto ester is known to chelate divalent metal ions, we speculate that compound **1** might potentially act as a multicopper oxidase inhibitor [37] to prohibit manganese oxide mineralization. Therefore, a maximum substrate concentration of 5.0 mM was selected for the following whole-cell biocatalytic oxidation of substrate **1**.

2.5. Continuous Live MnB1 Catalyzed Synthesis of α-Hydroxy-β-Keto Ester

As the whole-cell catalysts have been previously demonstrated as a continuous and repeated-batch reaction system [38], herein we explored such potential of live MnB1 for α-hydroxylation of **1**. The MnB1 bacteria were cultivated in Lept medium containing 1.0 mM $MnCl_2$ until 0.45 mM BMO were formed according to leucoberbelin blue (LBB) test [39]. The substrate **1** was then directly added to the culture medium to reach the concentration of 5.0 mM. During the first round of synthesis, the reaction progress was monitored as shown in Figure 4a. The reaction rate was fast at the initial stage but slowed down with the BMO being consumed. The highest conversion could reach 92% after 24 h (Figure 4b). Afterwards, the MnB1 bacteria were recycled by centrifugation, and washed by PBS before re-culturing in a fresh medium. The same concentration of fresh substrate was added to the medium when the BMO concentration reached 0.45 mM again. The yield of product **2** was consistently determined after 24 h. Notably, in consecutive four cycles, virtually identical yield was attainable for each run, and all were over 90% (92%, 91%, 93% and 94% respectively, Figure 4c). These results clearly suggest that the MnB1-based whole-cells catalyst can be easily regenerated and recycled for continuous reactions, which is highly valuable for low cost and sustainable catalysis and flow chemistry.

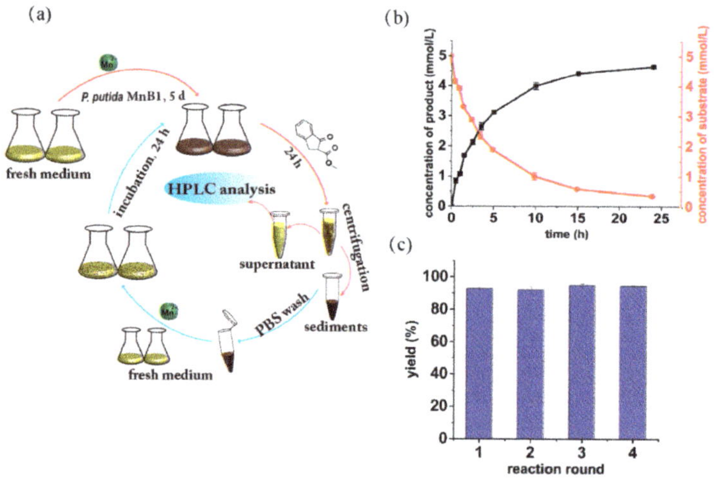

Figure 4. Synthesis of α-hydroxy-β-keto ester by the whole cells based on the biogenic manganese oxides: (**a**) Schematic diagram of the recycling experiments; (**b**) Profile of substrate consumption and product formation; (**c**) Yield of product **2** in recycling reaction.

3. Discussion

Manganese oxides are known to be generated via oxidation of Mn^{2+} through both abiotic and biotic channels in the environment. Related reports indicated that the abiotic manganese oxidation pathway is much slower than biological processes, which are performed by a large variety of bacteria and fungi [40,41]. Here, we used *P. putida* MnB1 as the model strain to generate the BMO. Notably, *P. putida* MnB1 is ubiquitous in the freshwater and soil, and its optimal growth temperature is 26~30 °C [20], which indicates that the strains are readily available and can be cultivated even in complicated environments. Thus, the robustness of *P. putida* MnB1 lays the groundwork for the prospective industrial application of BMO.

In recent years, BMO have received broad attention due to its marked redox reactivity, for instance, for the degradation of the organic pollutants, removal of heavy metal ions, etc. [42,43]. Although these processes involve various organic transformations, no rational synthesis of organic compounds with specific bond formation using BMO has been disclosed. In this study, we realized the synthesis of valuable α-hydroxy-β-keto ester **2** by utilizing BMO as an oxidant to promote the highly selective C-O bond formation event. Its superior performance than commercially available CMO highlights a strong capacity for synthetic oxidative reactions, and this is consistent with the previous studies on degradation applications [44]. In fact, oxidation of substrate **1** with BMO was a spot-to-spot process based on thin layer chromatography (TLC) analysis, and it was much faster and cleaner than that with CMO. Mechanistically, BMO produced by the *P. putida* MnB1 is known to contain larger inter-layer space and less structural Mn(III) than the CMO [45], thus providing larger specific surface areas [46]. We presume that these characteristics of BMO are beneficial to render defined interaction with substrates, and likely account for their better performance observed in our experiments.

Despite remarkable performance of isolated BMO for the oxidation of β-keto ester **1**, we were further motivated to develop an environmentally friendly and efficient catalytic system by using whole-cell bacteria as a novel type of catalyst. However, tremendous challenges need to be taken into consideration, such as the compatibility with reaction media, the toxicity of substrate to cells and others [47]. Results of solvent optimization (Figure 2a) guided us to take mixture of H_2O and MeCN as a suitable reaction medium. However, toxicity experiments showed that higher concentration (10 mM) of substrate would drastically inhibit the mineralization function of bacteria (Figure 3b). Jung [35] and Francis [34] independently concluded that the generation of BMO by *P. putida* MnB1 was fundamentally accomplished by multicopper oxidase, which catalytically oxidizes Mn(II) into Mn(VI) in cell extracts within its active center with three Cu(II) sites. This activity would get inhibited by heating or by treatment with a protease. The β-keto ester is known to chelate divalent metal ions, an essential component in the active site that interacts with amino acid residues of the enzyme. Therefore, this type of substrate might potentially act as a protein inhibitor [37]. We estimated that the higher concentration of β-keto ester **1** suppressed the manganese oxidizing partially due to the toxicity of substrate to the bacteria growth (Figure 3a and Figure S4), and this was probably a result of the impact of substrate on the activity of the multicopper oxidase. Based on these findings, we chose 5 mM as an appropriate concentration of substrate to conduct the whole-cell biocatalysis. As the content of β-keto ester **1** was 5-fold to Mn(II), the obtained 92% yield of product **2** clearly indicates Mn is operative in a catalytic manner, i.e., the regeneration of BMO by *P. putida* MnB1. Although a similar notion has been offered in the degradation of organic pollutants by Ko et al. [18], however, the recyclability of the bacteria for continuous catalysis has never been demonstrated.

In conclusion, we have realized the synthesis of valuable α-hydroxy-β-keto ester (**2**) by utilizing BMO as an oxidant to promote the highly selective C-O bond formation. Its superior performance than commercially available CMO highlights a strong capacity for synthetic oxidative reactions. Consecutive repeated-batch synthesis with recovered bacteria by the whole-cell catalytic system was achieved with consistently high levels of yield recorded [38]. Collectively, the whole cell catalytic *P. putida* MnB1 with biogenic manganese oxides is highly robust, even amenable to organic transformation with somewhat toxic substrates. As Mn(IV) is extensively utilized in oxidative organic reactions, we anticipate that this efficient biological system is promising for the benign synthesis of various bioactive substances as well as bulk fine chemicals to meet the increasing demand for sustainable chemistry.

4. Materials and Methods

4.1. Preparation of Freeze-Dried BMO

The bacterial strain *Pseudomonas putida* MnB1 [American Type Culture Collection (ATCC) no. 23483] was cultured in the Lept medium (0.5 g/L yeast extract (Ruji, Shanghai, China), 0.5 g/L Casamino Acids (Coolaber, Beijing, China), 5 mM D(+)-glucose (Macklin, Shanghai, China), 10 mM

HEPES (N-2-hydroxyethylpiperazine-N′-2-ethanesulfonic acid, pH 7.5, GBCBIO, Guangzhou, China), 0.48 mM $CaCl_2$ (Macklin), 0.83 mM $MgSO_4$ (Macklin), 3.7 μM $FeCl_3$ (Macklin), and 1 mL of trace element solution (10 mg/L $CuSO_4{\bullet}5H_2O$, 44 mg/L $ZnSO_4{\bullet}7H_2O$, 20 mg $CoCl_2{\bullet}6H_2O$ and 13 mg/L $Na_2MoO_4{\bullet}2H_2O$, (Macklin)) containing 1mM $MnCl_2$ (Aladdin, Shanghai, China) at 30 °C and shaken at 150 rpm for 5 d [48]. The suspensions were centrifuged at 8000× g for 20 m and the supernatant was discarded, sediments of bacteria and BMO were diluted three times with deionized water by means of centrifugation (20 m at 8000× g), then the mixture of precipitates were collected for freeze-drying to obtain the dried BMO sample.

4.2. Quantification of Freeze-Dried BMO by Inductive Coupled Plasma Optical Emission Spectrometry (ICP-OES)

The 25 μg dry m-MnB1 was treated with 200 μL 90% HNO_3 overnight and heated at 70 °C for 30 min, then was added H_2O_2 (30%, 200 μL). The resulting mixture was treated to evaporate at 200 °C for 1 h and then dissolved in deionized water to 4 mL solution to be measured by Inductive Coupled Plasma Optical Emission Spectrometry (ICP-OES 7000 Plus, ThermoFisher, Waltham, MA, USA) to quantify the amount of Mn^{2+}. ICP-OES combined with a water cross-flow nebulizer and Ar was run as the carrier gas; auxiliary energy flow, coolant flow, and nebulizer flow were set as 0.7 L/min, 13.00 L/min, and 0.7 L/min, respectively. The analyses were calibrated by gravimetric standards with different concentrations (2 ppm, 4 ppm, 8 ppm, 16 and 32 ppm, respectively) that were measured before sample quantification.

4.3. Synthesis of α-Hydroxy-β-Keto Ester by the Dry m-MnB1 and CMO

A reaction mixture (400 μL, MeCN: H_2O = 1:1) containing 50 mM β-keto ester **1**, 5.0 mg dry m-MnB1 or 1.75 mg CMO was shaken at 150 rpm, 30 °C for 3 h. A control experiment was conducted without manganese oxide under otherwise identical reaction conditions. All reaction mixtures were monitored by TLC (hexanes/ethyl acetate = 3:1). Then, the resulting samples were centrifuged at 8000× g for 5 m, supernatant was extracted with dichloromethane and evaporated under reduced pressure. The dried residues were purified by silica gel chromatography with hexanes/ethyl acetate (20:1–10:1) to get desired products for the NMR analysis. Finally, the purified samples were used as standards for the following HPLC detection.

4.3.1. Optimization of the Reaction Medium

The reaction mixtures (400 μL, MeCN: H_2O = 3:1, 1:1, 1:3, 1:6 or 1:9) containing 50 mM β-keto ester (1) (Preparation of this substrate followed a known protocol [49]), 5.0 mg dry m-MnB1 or 1.75 mg CMO (Aladdin) were shaken at 150 rpm, 30 °C for 3 h. The reaction samples were centrifuged at 8000× g for 5 m and supernatant was analyzed using HPLC to calculate the yield of oxidation product.

4.3.2. Assays of Different Dosage of Manganese Oxides on the Oxidation Reaction

Substrate **1** (0.02 mmol) and dry m-MnB1/CMO (Chemical manganese oxides, Innochem, Beijing, China) (0.01, 0.02, 0.03, 0.04, or 0.05 mmol) were added to test tubes and dissolved in MeCN (40 μL) and H_2O (360 μL). The reaction samples were stirred at 150 rpm, 30 °C for 1.5 h. Then, the reaction systems were centrifuged at 8000× g for 5 m and supernatants were analyzed using HPLC to obtain the yield of the oxidation product.

4.3.3. Kinetic Measurements of Dry m-MnB1 and CMO

To further compare the reaction efficiency of dry m-MnB1 and CMO, kinetic parameters were detected. The reaction mixtures (**1** (0.02 mmol), dried-BMO/CMO (0.02 mmol), MeCN (40 μL) and H_2O (360 μL) were stirred at 150 rpm, 30 °C, and the reaction progress was monitored by yield of the oxidation product.

4.4. Bioassays of Substrate β-Keto Ester Tolerance of Pseudomonas putida MnB1

The bacterial strain *Pseudomonas putida* MnB1 was precultured overnight in LB medium (10 g/L tryptone, 10 g/L NaCl, 5 g/L yeast extract) at 150 rpm, 30 °C and transferred into fresh Lept medium containing 1 mM $MnCl_2$ and various concentrations of **1** (0 mM, 5 mM, 10 mM, 15 and 20 mM, respectively). The colony forming units (CFU) were quantified by using the standard spread plate method [50] every 5 h until the generation of BMO. Simultaneously, the $OD_{600\ nm}$ level of bacteria was performed with microplate reader (3020-675, ThermoFisher).

4.5. Investigation of the Effects of Substrate on Manganese Mineralization

The oxidized Mn content generated in the suspension that was described in the Section 4.1 was measured at the indicated time points by using leucoberbelin blue (LBB, Sigma-Aldrich, St. Louis, MO, USA) method as illustrated by Krumbein [39]. With negligible modifications, bacteria culture (10 μL) was added to 50 μL LBB. After 15 m in the dark at room temperature, deposits were removed by centrifugation. $KMnO_4$ was used as the standard to confirm the absorbance at 620 nm of supernatant.

4.6. Continuous Biocatalytic Experiments by Whole Cells Based on BMO

The bacterial strain *Pseudomonas putida* MnB1 was cultured in the Lept medium containing 1 mM $MnCl_2$ at 150 rpm, 30 °C for 5 d to generate the BMO. The concentration of BMO was quantified by LBB method before adding the 5 mM substrate **1** to the medium. The mixture was then continued to shake at 150 rpm, 30 °C to get target product and the reaction progress was monitored by HPLC analysis. After the first-round reaction, the bacteria with BMO were collected by centrifugation and diluted three times with deionized water. The precipitates were added to fresh Lept medium containing 1 mM $MnCl_2$ to allow the continuous mineralization of manganese at 30 °C and 150 rpm. The mixture was used for the second-round reactions under comparable conditions when the concentration of BMO was similar with the first-round initial content. Four consecutive rounds of biocatalysis by whole-cell were conducted and yield of product in each round was measured. All procedures were performed under rigorous aseptic conditions.

4.7. Methods and Conditions of HPLC Analysis

For HPLC detection, all of the collected supernatant samples were filtered over a 0.22 μm filter (Millipore, Billerica, MA, USA) and analyzed by HPLC system containing a LC-UV 100 absorbance detector (Wufeng, Shanghai, China). The compounds were separated on a reverse phase HPLC C18 column (C18 250 × 4.6 mm, 5 μm, Shodex, Tokyo, Japan) at a constant flow rate of 1.0 mL/min and analysed by UV/Vis detection at 254 nm. Solvent A was deionized water containing 5% trifluoroacetic acid (TFA, Aladdin) and solvent B was acetonitrile containing 5% TFA. A gradient from 0% to 70% solvent B was applied from 0 to 5 min, the solvent B decreased 70% to 55% from 5 to 20 m, then dropped to 5% during 5 m. All analyses were calibrated by the absorption of substrate and product standards with different concentrations.

4.8. Statistical Analysis

All data were presented as means ± standard deviation of three replicates. Statistical analysis was completed using SPSS 16.0 (Statistical Package for the Social Sciences, IBM Corp., Armonk, NY, USA)

Supplementary Materials: Supplementary materials can be found at http://www.mdpi.com/1422-0067/21/5/1709/s1.

Author Contributions: Conceptualization, J.L., Y.W. and F.Z.; methodology, J.G., H.G. and J.L.; data curation, J.G. and J.L.; investigation, J.G. and H.G.; writing—original draft preparation, J.G.; writing—review and editing, Y.W. and F.Z.; supervision, Y.W. and F.Z.; funding acquisition, J.L. and Y.W. All authors have read and agreed to the published version of the manuscript.

Funding: This research was funded by the National Key R&D Program of China (2018YFA0903500), the China Postdoctoral Science Foundation funded project (2018M630849), the 1000 Young Talent Program of China, and Hubei Technological Innovation Project (2019ACA125).

Acknowledgments: We thank the Analytical and Testing Centre of HUST, Analytical and Testing Centre of School of Chemistry and Chemical Engineering (HUST), and Research Core Facilities for Life Sciences (HUST) for instrument support.

Conflicts of Interest: The authors declare no conflict of interest.

Abbreviations

BMO	Biogenic manganese oxide
CMO	Chemical manganese oxide
MeCN	Acetonitrile
CFU	Colony forming unit
ICP-OES	Inductive coupled plasma optical emission spectrometry
TFA	Trifluoroacetic acid

References

1. Sheldon, R.A.; Brady, D. Broadening the scope of biocatalysis in sustainable organic synthesis. *ChemSusChem* **2019**, *12*, 2859–2881. [CrossRef] [PubMed]
2. Son, E.J.; Lee, Y.W.; Ko, J.W.; Park, C.B. Amorphous Carbon Nitride as a Robust Photocatalyst for Biocatalytic Solar-To-Chemical Conversion. *ACS Sustain. Chem. Eng.* **2018**, *7*, 2545–2552. [CrossRef]
3. Yuan, H.; Liu, H.; Du, J.; Liu, K.; Wang, T.; Liu, L. Biocatalytic production of 2, 5-Furandicarboxylic acid: Recent advances and future perspectives. *Appl. Microbiol. Biot.* **2020**, *104*, 527–543. [CrossRef] [PubMed]
4. Sandoval, B.A.; Hyster, T.K. Emerging strategies for expanding the toolbox of enzymes in biocatalysis. *Curr. Opin. Chem. Biol.* **2020**, *55*, 45–51. [CrossRef]
5. Skellam, E. Strategies for engineering natural product biosynthesis in fungi. *Trends Biotechnol.* **2019**, *37*, 416–427. [CrossRef]
6. Woodley, J.M. New frontiers in biocatalysis for sustainable synthesis. *Curr. Opin. Green Sustain. Chem.* **2020**, *21*, 22–26. [CrossRef]
7. Akanuma, S.; Bessho, M.; Kimura, H.; Furukawa, R.; Yokobori, S.-I.; Yamagishi, A. Establishment of mesophilic-Like catalytic properties in a thermophilic enzyme without affecting its thermal stability. *Sci. Rep.* **2019**, *9*, 1–11. [CrossRef]
8. Hepworth, L.; France, S.; Hussain, S.; Both, P.; Turner, N.; Flitsch, S. De Novo Enzyme Cascades in Whole Cells for the Synthesis of Chiral Cyclic Amines. *ACS Catal.* **2017**, *4*, 2920–2925. [CrossRef]
9. Duetz, W.A.; Van Beilen, J.B.; Witholt, B. Using proteins in their natural environment: Potential and limitations of microbial whole-Cell hydroxylations in applied biocatalysis. *Curr. Opin. Biotechnol.* **2001**, *12*, 419–425. [CrossRef]
10. Liao, X.; Hou, J.; Wang, Y.; Zhang, H.; Sun, Y.; Li, X.; Tang, S.; Kato, K.; Yamauchi, M.; Jiang, Z. An active, selective, and stable manganese oxide-supported atomic Pd catalyst for aerobic oxidation of 5-hydroxymethylfurfural. *Green Chem.* **2019**, *21*. [CrossRef]
11. Miao, L.; Wang, J.; Zhang, P. Review on manganese dioxide for catalytic oxidation of airborne formaldehyde. *Appl. Surf. Sci.* **2019**, *466*, 441–453. [CrossRef]
12. Najafpour, M.M.; Renger, G.; Hołyńska, M.; Moghaddam, A.N.; Aro, E.-M.; Carpentier, R.; Nishihara, H.; Eaton-Rye, J.J.; Shen, J.-R.; Allakhverdiev, S.I. Manganese compounds as water-Oxidizing catalysts: From the natural water-Oxidizing complex to nanosized manganese oxide structures. *Chem. Rev.* **2016**, *116*, 2886–2936. [CrossRef] [PubMed]
13. Reidies, A.H. *Manganese Compounds*; Wiley-VCH Verlag GmbH & Co. KGaA: Weinheim, Germany, 2000.
14. Yu, Q.; Fu, Y.; Huang, J.; Qin, J.; Zuo, H.; Wu, Y.; Zhong, F. Enantioselective Oxidative Phenol-Indole [3+ 2] Coupling Enabled by Biomimetic Mn (III)/Brønsted Acid Relay Catalysis. *ACS Catal.* **2019**, *9*, 7285–7291. [CrossRef]
15. Jiang, L.; Huang, C.; Chen, J.; Chen, X. Oxidative transformation of 17β-Estradiol by MnO_2 in aqueous solution. *Arch. Environ. Contam. Toxicol.* **2009**, *57*, 221–229. [CrossRef] [PubMed]

16. Wang, G.; Liu, Y.; Wu, M.; Zong, W.; Yi, X.; Zhan, J.; Liu, L.; Zhou, H. Coupling the phenolic oxidation capacities of a bacterial consortium and in situ-generated manganese oxides in a moving bed biofilm reactor (MBBR). *Water Res.* **2019**, *166*, 115047. [CrossRef]
17. Zhang, H.; Huang, C.-H. Oxidative transformation of triclosan and chlorophene by manganese oxides. *Environ. Sci. Technol.* **2003**, *37*, 2421–2430. [CrossRef]
18. Tran, T.N.; Kim, D.-G.; Ko, S.-O. Synergistic effects of Biogenic Manganese Oxide and Mn(II)-Oxidizing bacterium *Pseudomonas putida* strain MnB1 on the degradation of 17 α-Ethinylestradiol. *J. Hazard. Mater.* **2018**, *344*, 350–359. [CrossRef]
19. Zhou, D.; Kim, D.-G.; Ko, S.-O. Heavy metal adsorption with biogenic manganese oxides generated by Pseudomonas putida strain MnB1. *J. Eng. Chem.* **2015**, *24*, 132–139. [CrossRef]
20. Caspi, R.; Tebo, B.M.; Haygood, M.G. c-Type Cytochromes and Manganese Oxidation in Pseudomonas putida MnB1. *Appl. Environ. Microbiol.* **1998**, *64*, 3549–3555. [CrossRef]
21. Martínez-Ruiz, E.B.; Cooper, M.; Fastner, J.; Szewzyk, U. Manganese-Oxidizing bacteria isolated from natural and technical systems remove cylindrospermopsin. *Chemosphere* **2020**, *238*, 124625. [CrossRef]
22. Su, J.; Deng, L.; Huang, L.; Guo, S.; Liu, F.; He, J. Catalytic oxidation of manganese (II) by multicopper oxidase CueO and characterization of the biogenic Mn oxide. *Water Res.* **2014**, *56*, 304–313. [CrossRef] [PubMed]
23. Wang, Y.-N.; Tsang, Y.F.; Wang, H.; Sun, Y.; Song, Y.; Pan, X.; Luo, S. Effective stabilization of arsenic in contaminated soils with biogenic manganese oxide (BMO) materials. *Environ. Pollut.* **2019**. [CrossRef] [PubMed]
24. Villalobos, M.; Lanson, B.; Manceau, A.; Toner, B.; Sposito, G. Structural model for the biogenic Mn oxide produced by Pseudomonas putida. *Am. Mineral.* **2006**, *91*, 489–502. [CrossRef]
25. Kimura, M. Recent topics in the syntheses of β-keto carboxylic acids and the derivatives. *Tetrahedron Lett.* **2018**, *59*, 1295–1300. [CrossRef]
26. Ren, Q.; Huang, J.; Wang, L.; Li, W.; Liu, H.; Jiang, X.; Wang, J. Highly Efficient Assembly of 3-Hydroxy Oxindole Scaffold via a Catalytic Decarboxylative [1,2]-Addition Strategy. *ACS Catal.* **2012**, *2*, 2622–2625. [CrossRef]
27. Yang, F.; Zhao, J.; Tang, X.; Wu, Y.; Yu, Z.; Meng, Q. Visible Light-Induced Salan-Copper (II)-Catalyzed Enantioselective Aerobic α-Hydroxylation of β-Keto Esters. *Adv. Syn. Catal.* **2019**, *361*, 1673–1677. [CrossRef]
28. Lu, M.; Zhu, D.; Lu, Y.; Zeng, X.; Tan, B.; Xu, Z.; Zhong, G. Chiral Brønsted acid-Catalyzed enantioselective α-Hydroxylation of β-Dicarbonyl compounds. *J. Am. Chem. Soc.* **2009**, *131*, 4562–4563. [CrossRef]
29. Naganawa, Y.; Aoyama, T.; Kato, K.; Nishiyama, H. Cu (II)-Catalyzed Enantioselective α-Hydroxylation and α-Chlorination of β-Ketoesters with N, N, O-Tridentate Chiral Phenanthroline Ligand. *Chem. Sel.* **2016**, *1*, 1938–1942.
30. Yang, F.; Zhao, J.; Tang, X.; Zhou, G.; Song, W.; Meng, Q. Enantioselective α-Hydroxylation by Modified Salen-Zirconium (IV)-Catalyzed Oxidation of β-Keto Esters. *Organic Lett.* **2017**, *19*, 448–451. [CrossRef]
31. Yin, X.P.; Zhu, L.; Zhou, J. Metal-Free Azidation of α-Hydroxy Esters and α-Hydroxy Ketones Using Azidotrimethylsilane. *Adv. Syn. Catal.* **2018**, *360*, 1116–1122. [CrossRef]
32. Zhang, Z.; Gao, X.; Yu, H.; Bi, J.; Zhang, G. Tandem Oxidative α-Hydroxylation/β-Acetalization Reaction of β-Ketoamides and Its Applications. *ACS Omega* **2017**, *2*, 7746–7754. [CrossRef] [PubMed]
33. Zou, L.; Wang, B.; Mu, H.; Zhang, H.; Song, Y.; Qu, J. Development of tartaric acid derived chiral guanidines and their application to catalytic enantioselective a-hydroxylation of b dicarbonyl compounds. *Org. Lett.* **2013**, *15*, 3106–3109. [CrossRef] [PubMed]
34. Francis, C.A.; Tebo, B.M. cumA Multicopper Oxidase Genes from Diverse Mn (II)-Oxidizing and Non-Mn (II)-OxidizingPseudomonas Strains. *Appl. Environ. Microbiol.* **2001**, *67*, 4272–4278. [CrossRef] [PubMed]
35. Jung, W.K.; Schweisfurth, R. Manganese oxidation by an intracellular protein of a *Pseudomonas* species. *J. Basic Microb.* **2007**, *19*, 107–115.
36. Soldatova, A.V.; Romano, C.A.; Tao, L.; Stich, T.A.; Casey, W.H.; Britt, R.D.; Tebo, B.M.; Spiro, T.G. Mn (II) oxidation by the multicopper oxidase complex Mnx: A coordinated two-Stage Mn (II)/(III) and Mn (III)/(IV) mechanism. *J. Am. Chem. Soc.* **2017**, *139*, 11381–11391. [CrossRef]
37. Grobler, J.A.; Stillmock, K.; Hu, B.; Witmer, M.; Felock, P.; Espeseth, A.S.; Wolfe, A.; Egbertson, M.; Bourgeois, M.; Melamed, J. Diketo acid inhibitor mechanism and HIV-1 integrase: Implications for metal binding in the active site of phosphotransferase enzymes. *Proc. Natl. Acad. Sci. USA* **2002**, *99*, 6661–6666. [CrossRef]

38. Ninh, P.H.; Honda, K.; Yokohigashi, Y.; Okano, K.; Omasa, T.; Ohtake, H. Development of a Continuous Bioconversion System Using a Thermophilic Whole-Cell Biocatalyst. *Appl. Environ. Microbiol.* **2013**, *79*, 1996–2001. [CrossRef]
39. Krumbein, W.E.; Altmann, H.J. A new method for the detection and enumeration of manganese oxidizing and reducing microorganisms. *Helgoländer Wiss. Meeresunters.* **1973**, *25*, 347–356. [CrossRef]
40. Hastings, D.; Emerson, S. Oxidation of manganese by spores of a marine bacillus: Kinetic and thermodynamic considerations. *Geochim. et Cosmochim. Acta* **1986**, *50*, 1819–1824. [CrossRef]
41. Das, A.P.; Sukla, L.B.; Pradhan, N.; Nayak, S. Manganese biomining: A review. *Bioresour. Technol.* **2011**, *102*, 7381–7387. [CrossRef]
42. Miyata, N.; Tani, Y.; Sakata, M.; Iwahori, K. Microbial manganese oxide formation and interaction with toxic metal ions. *J. Biosci. Bioeng.* **2007**, *104*, 1–8. [CrossRef] [PubMed]
43. Wang, R.; Wang, S.; Tai, Y.; Tao, R.; Dai, Y.; Guo, J.; Yang, Y.; Duan, S. Biogenic manganese oxides generated by green algae Desmodesmus sp. WR1 to improve bisphenol A removal. *J. Hazard. Mater.* **2017**, *339*, 310–319. [CrossRef] [PubMed]
44. Spiro, T.G.; Bargar, J.R.; Sposito, G.; Tebo, B.M. Bacteriogenic Manganese Oxides. *Acc. Chem. Res.* **2010**, *43*, 2–9. [CrossRef] [PubMed]
45. Liu, Y.; Li, Y.; Chen, N.; Ding, H.; Zhang, H.; Liu, F.; Yin, H.; Chu, S.; Wang, C.; Lu, A. Cu (II) sorption by biogenic birnessite produced by Pseudomonas putida strain MnB1: Structural differences from abiotic birnessite and its environmental implications. *Cryst. Eng. Comm.* **2018**, *20*, 1361–1374. [CrossRef]
46. Forrez, I.; Carballa, M.; Verbeken, K.; Vanhaecke, L.; Ternes, T.; Boon, N.; Verstraete, W. Diclofenac Oxidation by Biogenic Manganese Oxides. *Environ. Sci. Technol.* **2010**, *44*, 3449–3454. [CrossRef]
47. Florian, R. Whole-Cell based synthetic enzyme cascades-Light and shadow of a promising technology. *Curr. Opin. Chem. Biol.* **2019**, *49*, 84–90.
48. Boogerd, F.C.; de Vrind, J.P. Manganese oxidation by Leptothrix discophora. *J. Bacteriol.* **1987**, *169*, 489–494. [CrossRef]
49. Gu, X.; Zhang, Y.; Xu, Z.-J.; Che, C.-M. Iron (III)–Salan complexes catalysed highly enantioselective fluorination and hydroxylation of β-keto esters and N-Boc oxindoles. *Chem. Comm.* **2014**, *50*, 7870–7873. [CrossRef]
50. Boukouvalas, D.T.; Prates, R.A.; Leal, C.R.L.; de Araújo, S.A. Automatic segmentation method for CFU counting in single plate-Serial dilution. *Chemom. Intell. Lab. Syst.* **2019**, *195*, 103889. [CrossRef]

International Journal of
Molecular Sciences

MDPI

Article

Improving the Performance of Horseradish Peroxidase by Site-Directed Mutagenesis

Diana Humer and Oliver Spadiut *

TU Wien, Institute of Chemical, Environmental and Bioscience Engineering, Research Area Biochemical Engineering, Gumpendorfer Straße 1a, 1060 Vienna, Austria; diana.humer@tuwien.ac.at

* Correspondence: oliver.spadiut@tuwien.ac.at

Received: 23 January 2019; Accepted: 16 February 2019; Published: 20 February 2019

Abstract: Horseradish peroxidase (HRP) is an intensely studied enzyme with a wide range of commercial applications. Traditionally, HRP is extracted from plant; however, recombinant HRP (rHRP) production is a promising alternative. Here, non-glycosylated rHRP was produced in *Escherichia coli* as a DsbA fusion protein including a Dsb signal sequence for translocation to the periplasm and a His tag for purification. The missing *N*-glycosylation results in reduced catalytic activity and thermal stability, therefore enzyme engineering was used to improve these characteristics. The amino acids at four *N*-glycosylation sites, namely N13, N57, N255 and N268, were mutated by site-directed mutagenesis and combined to double, triple and quadruple enzyme variants. Subsequently, the rHRP fusion proteins were purified by immobilized metal affinity chromatography (IMAC) and biochemically characterized. We found that the quadruple mutant rHRP N13D/N57S/N255D/N268D showed 2-fold higher thermostability and 8-fold increased catalytic activity with 2,2′-azino-*bis*(3-ethylbenzothiazoline-6-sulphonic acid) (ABTS) as reducing substrate when compared to the non-mutated rHRP benchmark enzyme.

Keywords: *E. coli*; recombinant horseradish peroxidase; site-directed mutagenesis; periplasm; glycosylation sites

1. Introduction

The enzyme horseradish peroxidase (EC 1.11.1.7) is a heme-containing oxidoreductase used in both research and diagnostics for a wide range of applications (e.g., immunoassays, diagnostic kits, probe-based assay techniques as ELISA, EMSA, Western blotting and Southern blotting, waste water treatment and as a reagent in organic synthesis [1–7]. This 308 amino acid metalloenzyme incorporates two calcium atoms and four disulphide bridges [8,9]. In the plant, HRP also contains a hydrophobic 30 amino acid N-terminal leader sequence and a 15 amino acid C-terminal extension [10]. The C-terminal sequence is a sorting signal responsible for secretion to the vacuole [11] and the N-terminal sequence directs the enzyme to the endoplasmatic reticulum (ER) for post translational modifications, namely heme and calcium incorporation, disulphide bond formation and *N*-glycosylation [12]. The asparagine-linked glycans represent about a fifth of the 44 kDa plant holoenzyme. HRP possesses nine potential glycosylation sites and the pattern as well as the occupation of these sites is heterogeneous between HRP variants [13,14]. At least 28 different native HRP isoforms have been described so far [15] out of which HRP C is the most abundant and therefore the most studied one [9].

The commercially available HRP is extracted from *Armoracia rusticana* roots. However, only seasonal availability and long cultivation times paired with low yields make the classical production process undesirable. Moreover, the content of single isoenzymes is often very low and downstream processing is tedious. As a consequence, there is a need for a uniform enzyme preparation with defined characteristics and recombinant protein production would mitigate this issue. Many studies have addressed *Saccharomyces cerevisiae* or *Pichia pastoris* as host organisms for rHRP production because yeasts are easy to cultivate and commonly used for glycosylated and disulphide bond containing proteins. Alas, in yeast hyper-glycosylation occurs and the downstream process is cumbersome. Prokaryotes on the other hand lack the organelles necessary for glycosylation, namely ER and Golgi apparatus. In addition, the reducing environment in the cytoplasm of bacteria impedes the formation of disulphide bridges. Hence, recombinant glycoproteins with disulphide bridges are usually not produced in bacteria. However, *E. coli* is a very convenient host organism because of its cheap and easy cultivation at high cell densities. Moreover, there are no obstacles due to hyper-glycosylation as it is the case in yeast. This substantially facilitates downstream processing and allows application of the product for therapeutic use. It has also been shown that glycosylation is not essential for HRP activity or folding [16] although enzyme activity and thermal stability are considerably reduced when compared to the native enzyme [15,17]. Many studies have already been performed with the goal to enhance the general stability and activity properties of rHRP (Table 1).

Table 1. List of rHRP mutations that improve enzyme performance, listed by authors.

Mutation	Effect	Reference
N13D	Increased stability towards H_2O_2 Increased thermal stability	Asad et al. [18] Capone et al. [19]
N268D	Increased stability towards H_2O_2 Increased thermal stability Increased substrate specificity for phenol/4-aminoantipyrine Increased activity with phenol/4-aminoantipyrine	Asad et al. [18] Asad et al. [20] Capone et al. [19]
N268G	Increased stability towards H_2O_2 Increased thermal stability Increased substrate specificity for phenol/4-aminoantipyrine	Asad et al. [20]
N57S	Increased activity with ABTS Increased activity with H_2O_2 Increased thermal stability	Capone et al. [19]
N186D	Increased activity with ABTS	Capone et al. [19]
N198D	Increased substrate specificity for ABTS	Capone et al. [19]
N255D	Better folding in *E. coli* Increased activity with ABTS Increased activity with H_2O_2	Lin et al. [21] Capone et al. [19]
N158D	Increased activity with H_2O_2	Capone et al. [19]
K232N	Increased activity with ABTS Increased thermal stability Increased solvent stability Increased stability towards H_2O_2	Ryan et al. [22] Ryan et al. [23]
K232F	Increased activity with ABTS Increased thermal stability Increased solvent stability	Ryan et al. [22]
E238Q	Increased substrate specificity for ABTS	Ryan et al. [22]
K241N	Increased activity with ABTS	Ryan et al. [22]
K241E	Increased substrate specificity for ABTS Increased activity with ABTS	Ryan et al. [22]
K241A	Increased activity with ABTS Increased stability towards H_2O_2	Ryan et al. [22] Ryan et al. [23]

Table 1. *Cont.*

Mutation	Effect	Reference
K232N/K241N	Increased thermal stability Increased stability towards H_2O_2	Ryan et al. [22] Ryan et al. [23]
K232F/K241N	Increased activity with ABTS Increased thermal stability Increased solvent stability Increased stability towards H_2O_2	Ryan et al. [22] Ryan et al. [23]
K232Q/K241Q	Increased activity with ABTS	Ryan et al. [22]
T110V	Increased stability towards H_2O_2 Increased thermal stability	Ryan et al. [24] Ryan et al. [23]
T102A	Increased activity with ABTS	Ryan et al. [24]
K232E	Increased stability towards H_2O_2	Ryan et al. [23]
K241F	Increased stability towards H_2O_2	Ryan et al. [23]
R118K/R159K/K232N/K241F	Increased thermal stability Increased stability towards H_2O_2	Ryan et al. [25]
13A7 *	Increased activity with guaiacol	Morawski et al. [26]
H2-10G5 *	Increased activity with guaiacol Increased activity with ABTS Increased pH stability Increased thermal stability Increased stability towards SDS/urea/sodium chloride	Morawski et al. [26]
13A7-N175S *	Increased activity with guaiacol Increased activity with ABTS Increased pH stability Increased thermal stability Increased stability towards SDS/urea/sodium chloride	Morawski et al. [26]
13A10 *	Increased activity with guaiacol Increased activity with ABTS	Morawski et al. [27]
17E12 *	Increased activity with guaiacol Increased activity with ABTS	Morawski et al. [27]

* 13A7 = A85(GCC→GCT)/N212D/Q223L; * H2-10G5 = A85(GCC→GCT)/N175S/N212D; * 13A7-N175S = A85(GCC→GCT)/N212D/Q223L/N175S; * 13A10 = R93L/T102A/L131P/N135(AAC→AAT)/L223Q/T257 (ACT→ACA)/V303E; * 17E12 = N47S/T102A/G121(GGT→GGC)/L131P/N135(AAC→AAT)/L223Q/P226Q/T257 (ACT→ACA)/P289(CCT→CCA).

Lin et al. [21] identified a N255D mutant by random mutagenesis with 14-fold higher activity than the non-mutated benchmark enzyme but they concluded that this increase was due to better folding of the enzyme rather than improved catalytic performance. Directed evolution was used to identify mutants 13A10 and 17E12 (for mutant descriptions see Table 1, Footnotes) in *P. pastoris*, which were associated with increased specific activity with ABTS (5.4-fold and 2.8-fold) and guaiacol (2.4-fold and 1.2 fold) as substrates. The thermostability of 13A10 was comparable to the non-mutated benchmark enzyme but it was remarkably decreased in 17E12 mutants [27]. Variant 13A10 was used as starting point for successive rounds of directed evolution and gave rise to 13A7, H2-10G5 and 13A7-N175S (for mutant descriptions see Table 1, Footnotes). These variants were found to be more stable towards pH, temperature, SDS, urea and sodium chloride but enzyme activity was not further improved [26]. Ryan et al. [23] intensely studied the influence of site-directed mutations on hydrogen peroxide tolerance. They identified T110V, K232N and K241F, which were 25-, 18- and 12-fold more resistant towards hydrogen peroxide than the non-mutated benchmark enzyme. These variants also showed increased tolerance to heat and solvents. In addition, K232N as well as K241F displayed higher turnover numbers (k_{cat}) with ABTS as reducing substrate [22]. Asad et al. [18] changed the amino acids present at two *N*-glycosylation sites of rHRP by site-directed mutagenesis. They described variants N13D and

N268D, which showed increased catalytic efficiency with phenol/4-aminoantipyrine and were both more stable in terms of hydrogen peroxide and heat tolerance. A follow-up study identified N268G, which showed 18-fold higher resistance towards hydrogen peroxide and 2.5-fold higher thermal stability [20]. Capone et al. [19] performed a profound investigation of *N*-glycosylation mutants in *P. pastoris*, where asparagines at eight sites were replaced by aspartic acid, serine or glutamine. They showed that the positive influence of N13D and N268D on thermal stability is also valid for expression in yeast and that the variant identified by Lin et al. [21] was apparently slightly beneficial in terms of catalytic activity. Interestingly, a variant with mutations at all eight *N*-glycosylation sites showed substantially decreased activity and thermal stability.

At the moment, HRP is not used for in vivo medical applications, because the plant glycosylation pattern differs significantly from human glycoforms and therefore has immunogenic potential [28]. This can be circumvented by reducing the glycosylation pattern to mannose-type glycans, which can be achieved by adding an ER retention sequence. Unfortunately, these glycans lead to rapid clearance from circulation in humans [28]. Nevertheless, a combination of HRP and paracetamol or the plant hormone indole-3-acetic acid (IAA) was found to be medically active in targeted cancer treatment [29,30]. Although this cytotoxic effect has been known since the nineties [31,32], up to now, HRP is not considered suitable for therapeutic use. For this application, a recombinantly produced single isoform free of glycosylation with sufficient stability and activity would be desirable. In this study, we investigated the *N*-glycosylation mutants N13D, N57S, N255D, N268D and combinations thereof for thermal stability and catalytic efficiency with the substrate ABTS for rHRP expressed in *E. coli*. Soluble rHRP was preferred for mutant screenings because refolding of rHRP from *E. coli* inclusion bodies is a complex and cumbersome procedure which still has to be optimized. Therefore, we chose an expression system that leads to translocation of rHRP into the periplasm. The aim of this work was to improve the traits of non-glycosylated rHRP towards higher stability and catalytic efficiency to increase suitability for medical applications. Indeed, a promising rHRP *N*-glycosylation mutant was identified and biochemically characterized.

2. Results and Discussion

2.1. Protein Production

Recombinant HRP variants were produced and then translocated to the periplasm by the DsbA signal sequence. Soluble proteins were isolated from the periplasm and rHRP was purified to gain active, correctly folded enzyme. The imidazole concentration in the IMAC binding buffer was at the upper limit given by the column manufacturer (GE Healthcare, Chicago, IL, USA) to avoid unspecific interactions between *E. coli* host cell proteins and the stationary IMAC phase. Nevertheless, several impurities were visible on the SDS PAGE of the IMAC eluate (data not shown). Therefore, rHRP concentrations were calculated using an SDS-PAGE HRP standard curve with known concentrations (Figure S1). The peak area of rHRP was determined using Fiji Image Analysis Software (https://fiji.sc) [33] and the protein content was calculated using the slope of the linear regression line of known rHRP concentrations. This led to final enzyme titres of 0.05–0.09 mg rHRP g^{-1} DCW. The final rHRP product yield was between 0.04–0.08 g L^{-1} and is similar to reported values from Gundinger et al. [17] for soluble rHRP in pET39b$^+$ (0.048 g L^{-1}).

2.2. Biochemical Characterization

2.2.1. Biochemical Characterization of Benchmark rHRP and Seven rHRP Mutants

2.2.1.1. Enzyme Kinetics

Plant HRP VI-A (Sigma-Aldrich, St. Louis, MO, USA), non-mutated benchmark rHRP and the seven rHRP variants N13D, N57S, N255D, N268D, N57S/N268D, N57S/N255D/N268D and N13D/N57S/N255D/N268D were analysed for steady-state kinetics with ABTS as reducing substrate.

The kinetic constants are presented in Table 2. N13D and N255D showed less catalytic efficiency (K_{cat}/K_m) than the benchmark enzyme and for N13D this is in accordance with Capone et al. [19]. For N255D on the other hand, Capone et al. [19] observed almost the same catalytic activity compared to the benchmark enzyme (1.1-fold increase). N268D had a 2-fold increased turnover number (k_{cat}) when compared to the non-mutated rHRP and the same trend was shown by Asad et al. [20], where a 2.6-fold enhanced k_{cat} with phenol/4-aminoantipyrine was reported. The slightly enhanced catalytic efficiency of N57S, when compared to the benchmark rHRP (1.2-fold), is in accordance with Capone et al. [19] (1.4-fold). The triple mutant N57S/N255D/N268D reached a 3.2-fold higher turnover number than the benchmark rHRP and with a 10-fold increase, N13D/N57S/N255D/N268D showed the highest fold change of k_{cat} when compared to the benchmark rHRP. However, only N57S and N13D/N57S/N255D/N268D had an increased catalytic efficiency when compared to the non-mutated rHRP (1.2-fold and 2-fold, respectively). In general, the results were greatly affected by the unusually high Michaelis Menten constants (K_m) and the considerable standard deviations (see Section 2.2.2).

Table 2. Kinetic characteristics of plant HRP, rHRP and seven rHRP variants with ABTS as reducing substrate measured in 50 mM BisTris/HCl pH 7, 7% glycerol, 100 mM NaCl.

HRP variant	K_m [mM]	V_{max} [mol^{-1} L^{-1} × s]	K_{cat} [s^{-1}]	K_{cat}/K_m [mM^{-1} s^{-1}]
Benchmark rHRP	2.82 ± 1.52	$1.7 \times 10^{-6} \pm 4.3 \times 10^{-7}$	1.52 ± 0.38	0.54 ± 0.32
N13D	3.29 ± 0.33	$8.8 \times 10^{-7} \pm 4.0 \times 10^{-8}$	1.04 ± 0.05	0.32 ± 0.04
N57S	3.22 ± 0.59	$2.7 \times 10^{-6} \pm 2.6 \times 10^{-7}$	2.10 ± 0.19	0.64 ± 0.13
N255D	4.37 ± 0.86	$8.9 \times 10^{-7} \pm 1.0 \times 10^{-7}$	1.10 ± 0.12	0.24 ± 0.05
N268D	7.85 ± 4.98	$2.4 \times 10^{-6} \pm 1.0 \times 10^{-6}$	3.00 ± 1.25	0.38 ± 0.29
N57S/N268D	4.18 ± 3.55	$1.0 \times 10^{-6} \pm 3.9 \times 10^{-7}$	1.50 ± 0.58	0.36 ± 0.34
N57S/N255D/N268D	9.52 ± 6.89	$4.0 \times 10^{-6} \pm 2.0 \times 10^{-6}$	4.81 ± 2.37	0.51 ± 0.44
N13D/N57S/N255D/N268D	13.6 ± 6.63	$1.7 \times 10^{-5} \pm 5.6 \times 10^{-6}$	15.1 ± 5.00	1.11 ± 0.65
HRP Type VI−A	9.46 ± 5.18	$5.7 \times 10^{-3} \pm 1.8 \times 10^{-3}$	271 ± 87.4	28.7 ± 18.3

2.2.1.2. Thermal Stability

The thermal stability of plant HRP VI-A, non-mutated benchmark rHRP and the seven rHRP variants was determined at 60 °C (Table 3). N13D and N268D were found to enhance stability towards heat, which is in accordance with Asad et al. [18] and Capone et al. [19]. However, Capone et al. [19] also reported a positive effect of N57S, which could not be confirmed. Variant N255D seemed to have no effect on thermal stability. The double mutant N57S/N268D was similar to the rHRP benchmark enzyme concerning temperature susceptibility, so apparently the effect of the mutations is not additive. N268D had the highest benefit and was 3.6-fold more stable than the non-mutated rHRP, whereas the quadruple mutant N13D/N57S/N255D/N268D was 2.4-fold enhanced. Although N268D was the most thermostable variant, the quadruple mutant was chosen for further investigations, as this variant showed promising results for both catalytic activity and thermal stability.

Table 3. Half-life of plant HRP, rHRP and seven rHRP variants at 60 °C in 50 mM BisTris/HCl pH 7, 7% glycerol, 100 mM NaCl.

HRP Variant	$t_{1/2}$ at 60 °C
Benchmark rHRP	2 min 39 s ± 16 s
N13D	3 min 53 s ± 16 s
N57S	2 min 46 s ± 14 s
N255D	2 min 48 s ± 9 s
N268D	9 min 32 s ± 2 min 18 s
N57S/N268D	2 min 14 s ± 56 s
N57S/N255D/N268D	5 min 51 s ± 18 s
N13D/N57S/N255D/N268D	6 min 19 s ± 12 s
HRP Type VI-A	117 min ± 9 min 55 s

$t_{1/2}$ = half life.

2.2.2. Catalytic Activity of Plant HRP under Different Conditions

Interestingly, all obtained Michaelis Menten constants (Table 2) were much higher than previously reported for soluble rHRP variants [17] and this was also the case for commercially available plant HRP [17,34,35]. The substantial variability of the obtained Michaelis Menten constants may be a direct result of the high K_m values we observed due to fast reactions and high slopes, as variability increases with reaction velocity (see Table 2). We assumed that the measurement buffer was influencing the results of the kinetic measurements, as this was the only apparent difference to the previous procedure from Gundinger et al. [17]. Therefore, the catalytic activity of plant HRP was determined in both buffers: 50 mM BisTris/HCl pH 7, 7% glycerol, 100 mM NaCl and 50 mM KH_2PO_4 pH 6 (Table 4). The results indicated that the high Michaelis Menten constants, as well as the high standard deviations, are indeed buffer dependent. The K_m value was 6-fold higher in 50 mM BisTris/HCl pH 7, 7% glycerol, 100 mM NaCl buffer when compared to 50 mM KH_2PO_4 pH 6. As a consequence, the catalytic efficiency was 9-fold enhanced when the measurements were performed in potassium phosphate buffer. The kinetic parameters observed with 50 mM KH_2PO_4 pH 6 are in accordance with Gundinger et al. [17], as they reported a K_m-value of 1.75 mM and a V_{max}-value of 567 U mg^{-1}.

Table 4. Comparison of kinetic characteristics measured with plant HRP Type VI-A in two different buffers.

Buffer	K_m [mM]	V_{max} [mol^{-1} L^{-1} × s]	K_{cat} [s^{-1}]	K_{cat}/K_m [mM^{-1} s^{-1}]
Buffer 1	9.46 ± 5.18	5.7 × 10^{-3} ± 1.8×10^{-3}	272 ± 87.4	28.7 ± 18.3
Buffer 2	1.51 ± 0.15	7.9 × 10^{-3} ± 6.3 × 10^{-4}	378 ± 30.0	251 ± 32.2

Buffer 1, 50 mM BisTris/HCl pH 7, 7% glycerol, 100 mM NaCl; Buffer 2, 50 mM KH_2PO_4 pH 6.

However, at this point it was still unclear whether these results were obtained because of the buffer substance, the pH change or the additives in the BisTris buffer. Therefore, the specific activity of plant HRP with ABTS as reducing substrate was examined under different conditions (Figure 1). Apparently, the additives glycerol and sodium chloride have a negative influence, as the specific activity was 2-fold enhanced when they were omitted. The two buffers 50 mM KH_2PO_4 pH 7 and 50 mM BisTris/HCl pH 7 led to comparable results. The catalytic activity at pH 5 was 3-fold enhanced when compared to pH 7 when the measurements were conducted in potassium phosphate buffer. Based on these results, we decided to use 50 mM KH_2PO_4 pH 5 for all consecutive measurements (Section 2.2.3).

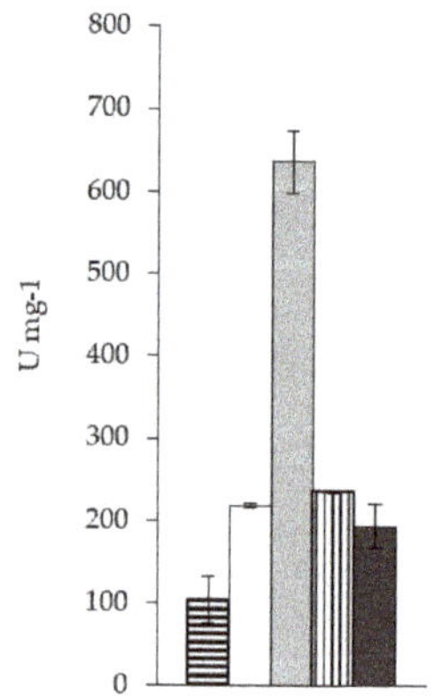

Figure 1. Specific activity of plant HRP Type VI-A with ABTS under various conditions. Five different buffers were used to measure the catalytic activity of plant HRP with 5 mM ABTS and 1 mM hydrogen peroxide. Horizontal stripes: 50 mM BisTris/HCl pH 7, 7% glycerol, 100 mM NaCl; white: 50 mM BisTris/HCl pH 7; light grey: 50 mM KH_2PO_4 pH 5; vertical stripes: 50 mM KH_2PO_4 pH 6.5; dark grey: 50 mM KH_2PO_4 pH 7.

2.2.3. Optimized Biochemical Characterization of Benchmark rHRP and Mutant N13D/N57S/N255D/N268D

The first investigation (Section 2.2.1) showed that the quadruple mutant N13D/N57S/N255D/N268D was the most promising variant when compared to the non-mutated benchmark rHRP (Tables 2 and 3). Therefore, a second protein purification and biochemical characterization with optimized assay conditions was performed to confirm these results.

2.2.3.1. Enzyme Activity

Enzyme kinetic measurements with plant HRP, non-mutated benchmark rHRP and rHRP variant N13D/N57S/N255D/N268D were performed in 50 mM KH_2PO_4 pH 5 using 96-well plates. Here, N13D/N57S/N255D/N268D showed an 8-fold enhanced catalytic efficiency (K_{cat}/K_m) and an 8-fold increased turnover number (k_{cat}) when compared to the benchmark enzyme (Table 5). The increase in k_{cat} for N13D/N57S/N255D/N268D was similar to the results obtained with the previous assay (Table 2, 10-fold). Besides, the K_m (0.27 mM) for plant HRP was similar to values reported in literature: 0.27 mM [35] and 0.11 mM [34]. However, the K_m-values observed for plant HRP in potassium phosphate buffer differed significantly between the two assays (Tables 4 and 5). We assume that this might be due to differences in pH, as the Michaelis Menten constant in 50 mM KH_2PO_4 pH 6 was 1.5 mM (Table 4) and Gundinger et al. [17] reported 1.75 mM in 50 mM KH_2PO_4 pH 6.5. Gilfoyle et al. [35] and Grigorenko et al. [34] used sodium phosphate/citrate buffer at pH 5 and sodium acetate buffer at pH 5, respectively, which resulted in K_m-values of 0.27 mM [35] and 0.11 mM [34] which are in accordance with 0.27 mM in potassium phosphate (Table 5).

Table 5. Kinetic characteristics of plant HRP, rHRP and N13D/N57S/N255D/N268D with ABTS as reducing substrate in 50 mM KH_2PO_4 pH 5.

HRP variant	K_m [mM]	V_{max} [mol^{-1} L^{-1} × s]	K_{cat} [s^{-1}]	K_{cat}/K_m [mM^{-1} s^{-1}]
Benchmark rHRP	0.44 ± 0.10	$2.0 \times 10^{-6} \pm 9.8 \times 10^{-8}$	2.24 ± 0.11	5.07 ± 1.16
N13D/N57S/N255D/N268D	0.45 ± 0.12	$1.8 \times 10^{-5} \pm 1.1 \times 10^{-6}$	17.4 ± 1.01	39.1 ± 10.5
HRP Type VI-A	0.27 ± 0.05	$8.8 \times 10^{-3} \pm 6.0 \times 10^{-4}$	422 ± 28.9	1,572 ± 306

2.2.3.2. Thermal Stability

The thermal stability of plant HRP VI-A, non-mutated benchmark rHRP and rHRP variant N13D/N57S/N255D/N268D was determined at 60 °C (Table 6). The measurement was performed at pH 7 to guarantee comparability with the previous assay (Section 2.2.1.2). Interestingly, the half-life of all HRP species measured in 50 mM KH_2PO_4 buffer pH 7 was slightly higher than the half-life in 50 mM BisTris/HCl pH 7, 7% glycerol, 100 mM NaCl (Table 4). Asad et al. [36] and Haifeng et al. [37] found that phosphate buffers can influence the thermostability of horseradish peroxidase. Although the samples were kept in 50 mM BisTris/HCl pH 7, 7% glycerol, 100 mM NaCl during heat exposition in both experiments, it might still be possible that potassium phosphate had a stabilizing effect during the activity measurement. In both experiments the plant HRP was far more stable than the rHRP variants (Tables 3 and 6) because glycans improve enzyme stability [18]. The thermostability of the quadruple mutant N13D/N57S/N255D/N268D was 2-fold higher than the stability of the benchmark rHRP enzyme (Table 6) which is in accordance with the previous assay (Table 3; 2.4-fold).

Table 6. Half-life of plant HRP, rHRP and N13D/N57S/N255D/N268D at 60 °C in 50 mM KH_2PO_4 pH 7.

HRP Variant	$t_{1/2}$ at 60 °C
Benchmark rHRP	3 min 29 s ± 1 s
N13D/N57S/N255D/N268D	7 min 41 s ± 31 s
HRP Type VI-A	133 min ± 1 min 20 s

Summarizing, the quadruple mutant showed a significantly augmented performance concerning thermal stability as well as catalytic activity with ABTS as substrate. Therefore, this variant is considered a good starting point for further enzyme engineering approaches.

3. Materials and Methods

3.1. Chemicals

Chemicals were purchased from Carl Roth (Karlsruhe, Germany) or AppliChem (Darmstadt, Germany). Plant HRP Type VI-A (Cat. No.: P6782) was purchased from Sigma-Aldrich (St. Louis, MO, USA), enzymes were purchased from New England Biolabs (Ipswich, MA, USA) or Thermo Fisher Scientific Inc. (Waltham, MA, USA), 2,2′-azino-*bis*(3-ethylbenzothiazoline-6-sulphonic acid) (ABTS) was purchased from Sigma-Aldrich or AMRESCO® biochemical (Solon, OH, USA).

3.2. Strains and Plasmids

The *hrp* gene coding for HRP variant C1A was codon-optimized for *E. coli* and obtained from GenSript USA Inc. (Piscataway, NJ, USA). HRP was produced as a His6-tagged recombinant protein from pET39b$^+$ in the *E. coli* strain BL21 (DE3) (Lucigen, Middleton, WI, USA). The plasmid pET39b+ (Novagen, San Diego, CA, USA) encodes a Dsb tag for export and periplasmic folding, so that a DsbA fusion protein with a HIS tag between the *dsbA* sequence and the *hrp* sequence is generated.

3.3. Strain Generation by Site-Directed Mutagenesis

The following plasmids were constructed with standard molecular cloning techniques [38]. Whole plasmid PCR was used to introduce mutations in the *hrp* gene by site-directed mutagenesis. The 7 kb fragment was amplified with the respective oligonucleotides to generate single, double, triple and quadruple mutations (Table 7). All oligonucleotides were purchased from Microsynth (Balgach, Switzerland). Each PCR reaction contained 1× Q5 Reaction Buffer, 200 µM dNTP Mix, 200 nM of both forward and reverse primer, 100 ng template vector DNA and 1 U Q5 High-Fidelity DNA Polymerase. The PCR products were purified with the Monarch PCR & DNA Cleanup Kit from New England Biolabs (NEB, Ipswich, MA, USA) and the template plasmid DNA was removed by FastDigest *Dpn*I (Thermo Scientific™, Waltham, MA, USA) digestion. 1 FDU (FastDigest unit, see Abbreviations) of *Dpn*I was added to the cleaned PCR products and incubated overnight at 37 °C. After heat inactivation at 80 °C for 20 min, the plasmids were transformed into BL21 (DE3). All DNA inserts of the recombinant plasmids were verified by DNA sequencing (Microsynth, Balgach, Switzerland).

Table 7. Oligonucleotide primers to mutate four Asn residues that act as *N*-glycosylation sites to either Asp or Ser.

N-site	Name	Sequence (5′→3′ Direction)
Benchmark rHRP	pET39b$^+$_hrp_fwd	GCGAATGCCCATGGATATGCAACTG
Benchmark rHRP	pET39b$^+$_hrp_rev	CCCGGGACTCGAGTTACGAGTT
N13	N13D_fwd2	CTGCCCGGATGTGAGCAACA
N13	N13D_rev2	CGGGCAGCTATTATCATAGAAGG
N57	N57S fwd	CTGCTGGACAGCACCACGTCC
N57	N57S rev	GTCCAGCAGGATACTTGCATCACAGCC
N255	N255D_fwd2	TTAGTTCCCCGGATGC
N255	N255D_rev2	CGGGGAACTAAACAGTTCT
N268	N268D fwd	GTTCGTTCATTTGCCGATTCGACCCAGA
N268	N268D rev	GGCAAATGAACGAACCAGCGGAATCG

The mutated sites are underlined. fwd: forward; rev: reverse.

3.4. Growth Conditions and Protein Production

LB medium (10 g L^{-1} tryptone, 10 g L^{-1} NaCl and 5 g L^{-1} yeast extract) or SB medium (32 g L^{-1} tryptone; 20 g L^{-1} yeast extract; 5 g L^{-1} NaCl; 5 mM NaOH) was used for cultivation of BL21 (DE3) strains. Kanamycin was added to a final concentration of 50 mg L^{-1} to ensure plasmid maintenance. Pre-cultures were grown overnight at 37 °C with shaking (250 rpm) in 50 mL LB^{Kan} or SB^{Kan} medium and 2.5 L Ultra Yield Flasks (UYF) were inoculated to reach an optical density (OD_{600}) of 0.3 in a final volume of 500 mL LB^{Kan} or SB^{Kan} medium. The cells were grown at 37 °C with shaking (250 rpm) until an OD_{600} of 0.5, subsequently *hrp* expression was induced by adding 0.1 mM isopropyl β-D-1-thiogalactopyranoside (IPTG). After growth for 20 h at 25 °C and 250 rpm, the cells were harvested by centrifugation (4500 rpm, 30 min, 4 °C).

3.5. Protein Purification

The cell pellets were resuspended in buffer A (50 mM BisTris/HCl pH 7, 500 mM NaCl, 40 mM imidazole) with cOmplete™ Protease Inhibitor Cocktail (Roche, Basel, Switzerland). The cell suspension was homogenized with an Avestin Emulsiflex C3 high pressure homogenizer (Avestin, Ottawa, ON, Canada) for 10 passages at 1000 bar and centrifuged afterwards at 10,000 rpm for 1 h. Particles were removed from the supernatant by filtration (0.2 µm) prior to protein purification with IMAC using the Äkta pure system (GE Healthcare, Chicago, IL, USA). The column (HisTrap™ Fast Flow Crude 1 mL, GE Healthcare) was equilibrated with 10 column volumes (CV) buffer A and the crude extract was loaded with a linear flow rate of 156 cm h^{-1}. Subsequently, the column was washed with 10–20 CV buffer A before step elution with 100% buffer B (50 mM BisTris/HCl pH 7, 500 mM NaCl, 500 mM imidazole) at a linear flow rate of 156 cm h^{-1}. The eluted protein fractions were desalted with Sephadex G-25 PD-10 desalting columns (GE Healthcare, Chicago, IL, USA) and eluted in buffer C (50 mM BisTris/HCl pH 7, 7% glycerol, 100 mM NaCl). Glycerol and sodium chloride were added to the buffer as these substances were found to positively impact on rHRP stability (data not shown). Total protein content was measured using the Bradford assay [39] and the eluate was analysed by SDS PAGE. The concentration of rHRP was calculated as µM mL^{-1} using 61.5 kDa (with DsbA Protein and His tag) as molecular mass of the fusion protein. Hemin and calcium chloride were added in a 2-fold and 4-fold molar amount, respectively and the enzyme preparations were incubated over night at 4 °C with slight agitation.

3.6. SDS PAGE

SDS PAGE was performed according to the Laemmli protocol [40]. Mini-PROTEAN® TGX Stain-Free™ Precast Gels (Bio-Rad, Hercules, CA, USA) were used and the gel was run with SDS running buffer (25 mM Tris, 200 mM glycine, 0.1% SDS) in a Bio-Rad Mini-PROTEAN® Tetra Cell. Proteins were separated at 125 V for 1 h and the bands were visualized with Coomassie Brilliant Blue solution. Bio-Rad Precision Plus Protein™ Dual Xtra Prestained Protein Standard or Thermo Scientific PageRuler™ Plus Prestained Protein Ladder, 10 to 250 kDa (Waltham, MA, USA) were used as mass standards. The SDS PAGE was analysed using a ChemiDoc™ MP System with Image Lab™ Software (Bio-Rad, Hercules, CA, USA).

3.7. Biochemical Enzyme Characterization

3.7.1. Assay Development

3.7.1.1. Pathlength Correction for Microplates

The pathlength for 200 µL reaction volume was determined experimentally by measuring the absorbance of MilliQ water in the Tecan Infinite M200 PRO (Tecan, Männedorf, Switzerland) and the Hitachi U-2900 spectrophotometer (Hitachi, Tokyo, Japan). Water absorption can be measured at near infrared wavelength with a maximum absorbance at 975 nm, the measurement at 900 nm subtracts background absorbance (e.g., plastic of 96-well plate). The correct pathlength for the 96-well plates was then calculated according to Equation (1):

$$pathlength = \frac{A975\text{ nm }(well) - A900\text{ nm }(well)}{A975\text{ nm }(cuvette) - A900\text{ nm }(cuvette)} \times 10\text{ mm} \quad (1)$$

Thus, a reaction volume of 200 µL results in a pathlength of 0.58 cm in 96-well plates.

3.7.1.2. Determination of Extinction Coefficient for ABTS

The extinction coefficient of the ABTS radical was determined experimentally with a Tecan Infinite M200 PRO system (Tecan, Männedorf, Switzerland) for ABTS purchased from Sigma-Aldrich (St. Louis, MO, USA) and AMRESCO® biochemical (Solon, OH, USA). ABTS concentrations ranging from 0 mM to 0.15 mM were measured in 50 mM KH_2PO_4 pH 5 buffer with 1 mM hydrogen peroxide in a final volume of 200 µL and 22.5 ng plant HRP Type VI-A (Sigma-Aldrich, Cat. No.: P6782). The reaction was followed to the plateau phase in a 96-well plate at 420 nm for 40 min at 30 °C. Finally, the extinction coefficient was calculated from the slope of the linear regression line of ABTS concentration plotted against maximal absorbance. The extinction coefficient of ABTS at 420 nm is 27 mM^{-1} cm^{-1} for ABTS purchased from both companies (Sigma-Aldrich: y = 15.455x + 0.0548; R^2 = 0.9977/ AMRESCO® biochemical: y = 15.455x + 0.0314; R^2 = 0.9986).

3.7.1.3. Enzyme Activity Calculation

HRP activity in Units mL^{-1} (1 Unit is defined as the amount of enzyme which oxidizes 1 µmol of ABTS per minute) was calculated according to Equation (2):

$$\frac{U}{\text{mL}} = \frac{(\Delta A420\text{ nm} - \Delta A420\text{ nm } blank) \times total\ volume\ of\ reaction \times d}{\varepsilon \times \text{p} \times \text{sample volume}} \quad (2)$$

d = dilution factor
p = pathlength in cm, which is 1 for cuvettes and 0.58 in a 96-well plate when the reaction volume is 200 µL (see Section 3.7.1.1).
ε = 27 mM^{-1} cm^{-1}

3.7.2. Catalytic Activity of Plant HRP under Different Conditions

Enzyme activity of plant HRP Type VI-A (Sigma-Aldrich, Cat. No.: P6782, St. Louis, MO, USA) was determined either in 50 mM BisTris/HCl pH 7, 7% glycerol, 100 mM NaCl or in 50 mM KH_2PO_4 pH 6. The reaction mixture in the cuvette contained 27.6 ng HRP, a saturating hydrogen peroxide concentration of 1 mM and varying ABTS concentrations (0.1–10 mM) in 50 mM BisTris/HCl pH 7, 7% glycerol, 100 mM NaCl or in 50 mM KH_2PO_4 pH 6 with a final volume of 1 mL. The increase in absorption was followed at 420 nm for 180 s at 30 °C in a Hitachi U-2900 spectrophotometer (Hitachi, Tokyo, Japan). All measurements were performed in triplicates. Kinetic parameters were calculated using OriginPro software (OriginLab Corporation 2016, Northampton, MA, USA). As these measurements revealed substantial differences concerning the kinetic parameters between the two buffers, it was decided to further investigate the influence of buffer substance, pH and buffer additives in a 96-well plate format. Here, the reaction mixture in each well contained a saturating hydrogen peroxide concentration of 1 mM and 5 mM ABTS in a final volume of 200 µL. Commercially available plant HRP (9 ng; Sigma-Aldrich, Cat. No.: P6782, St. Louis, MO, USA) was added to the reaction mixture and the increase in absorption was followed at 420 nm for 180 s at 30 °C in a Tecan Infinite M200 PRO instrument (Tecan, Männedorf, Switzerland). The following buffers were used: 50 mM KH_2PO_4 pH 5; 50 mM KH_2PO_4 pH 6.5; 50 mM KH_2PO_4 pH 7; 50 mM BisTris/HCl pH 7, 7% glycerol, 100 mM NaCl or 50 mM BisTris/HCl pH 7. All measurements were performed in duplicates.

3.7.3. Enzyme Kinetics

Enzyme kinetic parameters for the substrate ABTS were determined with non-mutated rHRP benchmark enzyme, seven rHRP mutants and commercially available plant HRP Type VI-A. The samples were either measured in cuvettes with a U-2900 spectrophotometer (Hitachi, Tokyo, Japan) or in 96-well plates using a Tecan Infinite M200 PRO instrument (Tecan, Männedorf, Switzerland).

3.7.3.1. Hitachi U-2900 Spectrophotometer

The reaction mixture in the cuvette contained 20 μL protein sample, a saturating hydrogen peroxide concentration of 1 mM and varying ABTS concentrations (0.1–10 mM) in 50 mM BisTris/HCl pH 7, 7% glycerol, 100 mM NaCl buffer with a final volume of 1 mL. The increase in absorption was followed at 420 nm for 180 s at 30 °C in a Hitachi U-2900 spectrophotometer. All measurements were performed in triplicates. The kinetic parameters were calculated using OriginPro software (OriginLab Corporation 2016, Northampton, MA, USA).

3.7.3.2. Tecan Infinite M200 PRO

The reaction mixture in each well of the 96-well plate contained a saturating hydrogen peroxide concentration of 1 mM and varying ABTS concentrations (0.1–10 mM) in 50 mM KH_2PO_4 pH 5 buffer in a final volume of 200 μL. Protein sample (5 μL) was added to each well and was filled up with 195 μL reaction mixture. The increase in absorption was followed in a 96-well plate at 420 nm for 180 s at 30 °C in a Tecan Infinite M200 PRO instrument. Eight replicates were used for all measurements. The kinetic parameters were calculated using OriginPro software (OriginLab Corporation 2016, Northampton, MA, USA).

3.7.4. Thermal Stability

The thermal stability of the enzyme was assessed at 60 °C in 50 mM BisTris/HCl pH 7, 7% glycerol, 100 mM NaCl and the residual activity with ABTS was measured after 0, 1, 2, 3 and 5 min for the non-mutated rHRP benchmark enzyme and the seven rHRP mutants. The residual activity of plant HRP Type VI-A (Sigma-Aldrich, Cat. No.: P6782, St. Louis, MO, USA) was measured after 0, 30, 60, 90 and 120 min, respectively. The protein concentration was normalized to 0.05 g L^{-1} for all samples to minimize potential influences of protein concentrations on thermal stability. The reaction mixture contained 20 μL of protein, a saturating hydrogen peroxide concentration of 1 mM and 10 mM ABTS in 50 mM BisTris/HCl pH 7, 7% glycerol, 100 mM NaCl or in 50 mM KH_2PO_4 pH 7 with a final volume of 1 mL. The increase in absorption was followed at 420 nm for 180 s at 30 °C in a Hitachi U-2900 spectrophotometer (Hitachi, Tokyo, Japan). The residual enzyme activity was plotted against incubation time and the half-life at 60 °C was calculated using the rate of inactivation in Equation (3):

$$t_{1/2} = \frac{ln2}{k_{in}}$$

$t_{1/2}$ = half life
k_{in} = slope of the logarithmic residual activity

4. Conclusions

HRP has many features that make it suitable for therapeutic use: it is stable at 37 °C, shows high activity at physiological pH and can be conjugated to antibodies or lectins [9]. Thus, it is highly interesting to engineer rHRP for increased activity and stability for use in medical applications. In our study we discovered a new non-glycosylated rHRP variant with improved characteristics by site-directed mutagenesis of amino acids at the *N*-glycosylation sites. N13D/N57S/N255D/N268D was found to substantially increase activity with ABTS as substrate, the catalytic efficiency of this variant was 8-fold higher when compared to the rHRP benchmark enzyme. Moreover, N13D/N57S/N255D/N268D is

2-fold more stable towards high temperature exposure than the non-mutated rHRP. Currently, we work on further improvement of N13D/N57S/N255D/N268D by directed evolution, as well as the additional introduction of selected mutants (see Table 1). Finally, our future goal will be to produce the resulting non-glycosylated rHRP variant in a large-scale inclusion body process.

Supplementary Materials: Supplementary materials can be found at http://www.mdpi.com/1422-0067/20/4/916/s1.

Author Contributions: Conceptualization, O.S.; Methodology, O.S. and D.H.; Validation, O.S. and D.H.; Formal analysis, D.H.; Investigation, D.H.; Resources, O.S.; Data curation, D.H.; Writing—original draft preparation, D.H.; Writing—review and editing, O.S. and D.H.; Visualization, D.H.; Supervision, O.S.; Project administration, O.S.; Funding acquisition, O.S.

Funding: This research was funded by Open Access Funding by the Austrian Science Fund (FWF), grant number P30872-B26.

Conflicts of Interest: The authors declare no conflict of interest.

Abbreviations

ABTS	2,2′-azino-*bis*(3-ethylbenzothiazoline-6-sulphonic acid)
BisTris	*Bis*(2-hydroxyethyl)amino-tris(hydroxymethyl)methane
CV	Column volumes
DCW	Dry cell weight [g L^{-1}]
ELISA	Enzyme-linked Immunosorbent Assay
EMSA	Electrophoretic Mobility Shift Assay
ER	Endoplasmatic reticulum
FDU	FastDigest unit, 1 μL of enzyme (1 FDU) cleaves 1 μg of DNA substrate in 5–15 min at 37 °C in 20 μL of 1× FastDigest buffer
HRP	Plant horseradish peroxidase
IAA	Indole-3-acetic acid
IMAC	Immobilized Metal Affinity Chromatography
IPTG	Isopropyl β-D-1-thiogalactopyranoside
rHRP	Recombinant Horseradish Peroxidase
SDS	Sodium dodecyl sulfate
UYF	Ultra Yield™

References

1. Gholami-Borujeni, F.; Mahvi, A.H.; Naseri, S.; Faramarzi, M.A.; Nabizadeh, R.; Alimohammadi, M. Application of immobilized horseradish peroxidase for removal and detoxification of azo dye from aqueous solution. *Res. J. Chem. Environ.* **2011**, *15*, 217–222.
2. Krieg, R.; Halbhuber, K. Recent advances in catalytic peroxidase histochemistry. *Cell. Mol. Biol. (Noisy-le-Grand)* **2003**, *49*, 547–563.
3. Litescu, S.C.; Eremia, S.; Radu, G.L. Biosensors for the determination of phenolic metabolites. In *Bio-Farms for Nutraceuticals*; Springer: Boston, MA, USA, 2010; pp. 234–240. [CrossRef]
4. Marquette, C.A.; Blum, L.J. Chemiluminescent enzyme immunoassays: A review of bioanalytical applications. *Bioanalysis* **2009**, *1*, 1259–1269. [CrossRef] [PubMed]
5. Ryan, B.J.; Carolan, N.; O'Fagain, C. Horseradish and soybean peroxidases: Comparable tools for alternative niches? *Trends Biotechnol.* **2006**, *24*, 355–363. [CrossRef] [PubMed]
6. Vasileva, N.; Godjevargova, T.; Ivanova, D.; Gabrovska, K. Application of immobilized horseradish peroxidase onto modified acrylonitrile copolymer membrane in removing of phenol from water. *Int. J. Biol. Macromol.* **2009**, *44*, 190–194. [CrossRef] [PubMed]
7. Yang, H. Enzyme-based ultrasensitive electrochemical biosensors. *Curr. Opin. Chem. Biol.* **2012**, *16*, 422–428. [CrossRef] [PubMed]
8. Welinder, K.G. Covalent structure of the glycoprotein horseradish peroxidase (EC 1.11. 1.7). *FEBS Lett.* **1976**, *72*, 19–23. [CrossRef]

9. Veitch, N.C. Horseradish peroxidase: A modern view of a classic enzyme. *Phytochemistry* **2004**, *65*, 249–259. [CrossRef] [PubMed]
10. Huddy, S.M.; Hitzeroth, I.I.; Meyers, A.E.; Weber, B.; Rybicki, E.P. Transient Expression and Purification of Horseradish Peroxidase C in *Nicotiana benthamiana*. *Int. J. Mol. Sci.* **2018**, *19*, 115. [CrossRef] [PubMed]
11. Matsui, T.; Tabayashi, A.; Iwano, M.; Shinmyo, A.; Kato, K.; Nakayama, H. Activity of the C-terminal-dependent vacuolar sorting signal of horseradish peroxidase C1a is enhanced by its secondary structure. *Plant Cell Physiol.* **2011**, *52*, 413–420. [CrossRef]
12. Matsui, T.; Nakayama, H.; Yoshida, K.; Shinmyo, A. Vesicular transport route of horseradish C1a peroxidase is regulated by N-and C-terminal propeptides in tobacco cells. *Appl. Microbiol. Biotechnol.* **2003**, *62*, 517–522. [CrossRef] [PubMed]
13. Wuhrer, M.; Hokke, C.H.; Deelder, A.M. Glycopeptide analysis by matrix-assisted laser desorption/ionization tandem time-of-flight mass spectrometry reveals novel features of horseradish peroxidase glycosylation. *Rapid Commun. Mass Spectrom.* **2004**, *18*, 1741–1748. [CrossRef] [PubMed]
14. Wuhrer, M.; Balog, C.I.; Koeleman, C.A.; Deelder, A.M.; Hokke, C.H. New features of site-specific horseradish peroxidase (HRP) glycosylation uncovered by nano-LC-MS with repeated ion-isolation/fragmentation cycles. *Biochim. Biophys. Acta (BBA)-Gen. Subj.* **2005**, *1723*, 229–239. [CrossRef] [PubMed]
15. Krainer, F.W.; Glieder, A. An updated view on horseradish peroxidases: Recombinant production and biotechnological applications. *Appl. Microbiol. Biotechnol.* **2015**, *99*, 1611–1625. [CrossRef] [PubMed]
16. Smith, A.T.; Santama, N.; Dacey, S.; Edwards, M.; Bray, R.C.; Thorneley, R.; Burke, J.F. Expression of a synthetic gene for horseradish peroxidase C in *Escherichia coli* and folding and activation of the recombinant enzyme with Ca^{2+} and heme. *J. Biol. Chem.* **1990**, *265*, 13335–13343. [PubMed]
17. Gundinger, T.; Spadiut, O. A comparative approach to recombinantly produce the plant enzyme horseradish peroxidase in *Escherichia coli*. *J. Biotechnol.* **2017**, *248*, 15–24. [CrossRef] [PubMed]
18. Asad, S.; Khajeh, K.; Ghaemi, N. Investigating the structural and functional effects of mutating Asn glycosylation sites of horseradish peroxidase to Asp. *Appl. Biochem. Biotechnol.* **2011**, *164*, 454–463. [CrossRef]
19. Capone, S.; Pletzenauer, R.; Maresch, D.; Metzger, K.; Altmann, F.; Herwig, C.; Spadiut, O. Glyco-variant library of the versatile enzyme horseradish peroxidase. *Glycobiology* **2014**, *24*, 852–863. [CrossRef]
20. Asad, S.; Dastgheib, S.M.; Khajeh, K. Construction of a horseradish peroxidase resistant toward hydrogen peroxide by saturation mutagenesis. *Biotechnol. Appl. Biochem.* **2016**, *63*, 789–794. [CrossRef]
21. Lin, Z.; Thorsen, T.; Arnold, F.H. Functional expression of horseradish peroxidase in *E. coli* by directed evolution. *Biotechnol. Prog.* **1999**, *15*, 467–471. [CrossRef]
22. Ryan, B.J.; Ó'Fágáin, C. Effects of mutations in the helix G region of horseradish peroxidase. *Biochimie* **2008**, *90*, 1414–1421. [CrossRef] [PubMed]
23. Ryan, B.J.; Ó'Fágáin, C. Effects of single mutations on the stability of horseradish peroxidase to hydrogen peroxide. *Biochimie* **2007**, *89*, 1029–1032. [CrossRef] [PubMed]
24. Ryan, B.J.; O'Connell, M.J.; Ó'Fágáin, C. Consensus mutagenesis reveals that non-helical regions influence thermal stability of horseradish peroxidase. *Biochimie* **2008**, *90*, 1389–1396. [CrossRef] [PubMed]
25. Ryan, B.J.; Ó'Fágáin, C. Arginine-to-lysine substitutions influence recombinant horseradish peroxidase stability and immobilisation effectiveness. *BMC Biotechnol.* **2007**, *7*, 86. [CrossRef] [PubMed]
26. Morawski, B.; Quan, S.; Arnold, F.H. Functional expression and stabilization of horseradish peroxidase by directed evolution in *Saccharomyces cerevisiae*. *Biotechnol. Bioeng.* **2001**, *76*, 99–107. [CrossRef] [PubMed]
27. Morawski, B.; Lin, Z.; Cirino, P.; Joo, H.; Bandara, G., Arnold, F.H Functional expression of horseradish peroxidase in *Saccharomyces cerevisiae* and *Pichia pastoris*. *Protein Eng.* **2000**, *13*, 377–384. [CrossRef] [PubMed]
28. Brooks, S.A. Appropriate glycosylation of recombinant proteins for human use. *Mol. Biotechnol.* **2004**, *28*, 241–255. [CrossRef]
29. Tupper, J.; Tozer, G.M.; Dachs, G.U. Use of horseradish peroxidase for gene-directed enzyme prodrug therapy with paracetamol. *Br. J. Cancer* **2004**, *90*, 1858–1862. [CrossRef]
30. Dai, M.; Liu, J.; Chen, D.E.; Rao, Y.; Tang, Z.J.; Ho, W.Z.; Dong, C.Y. Tumor-targeted gene therapy using Adv-AFP-HRPC/IAA prodrug system suppresses growth of hepatoma xenografted in mice. *Cancer Gene Ther.* **2012**, *19*, 77–83. [CrossRef]
31. Folkes, L.; Candeias, L.; Wardman, P. Toward targeted "oxidation therapy" of cancer: Peroxidase-catalysed cytotoxicity of indole-3-acetic acids. *Int. J. Radiat. Oncol. Biol. Phys.* **1998**, *42*, 917–920. [CrossRef]

32. Folkes, L.K.; Wardman, P. Oxidative activation of indole-3-acetic acids to cytotoxic species—A potential new role for plant auxins in cancer therapy. *Biochem. Pharmacol.* **2001**, *61*, 129–136. [CrossRef]
33. Schindelin, J.; Arganda-Carreras, I.; Frise, E.; Kaynig, V.; Longair, M.; Pietzsch, T.; Preibisch, S.; Rueden, C.; Saalfeld, S.; Schmid, B. Fiji: An open-source platform for biological-image analysis. *Nat. Methods* **2012**, *9*, 676–682. [CrossRef] [PubMed]
34. Grigorenko, V.; Chubar, T.; Kapeliuch, Y.; Börchers, T.; Spener, F.; Egorova, A. New approaches for functional expression of recombinant horseradish peroxidase C in *Escherichia coli*. *Biocatal. Biotransform.* **1999**, *17*, 359–379. [CrossRef]
35. Gilfoyle, D.J.; Rodriguez-Lopez, J.N.; Smith, A.T. Probing the aromatic-donor-binding site of horseradish peroxidase using site-directed mutagenesis and the suicide substrate phenylhydrazine. *Eur. J. Biochem.* **1996**, *236*, 714–722. [CrossRef] [PubMed]
36. Asad, S.; Torabi, S.-F.; Fathi-Roudsari, M.; Ghaemi, N.; Khajeh, K. Phosphate buffer effects on thermal stability and H_2O_2-resistance of horseradish peroxidase. *Int. J. Biol. Macromol.* **2011**, *48*, 566–570. [CrossRef] [PubMed]
37. Haifeng, L.; Yuwen, L.; Xiaomin, C.; Zhiyong, W.; Cunxin, W. Effects of sodium phosphate buffer on horseradish peroxidase thermal stability. *J. Therm. Anal. Calorim.* **2008**, *93*, 569–574. [CrossRef]
38. Sambrook, J.; Fritsch, E.F.; Maniatis, T. *Molecular Cloning: A Laboratory Manual*; Cold Spring Harbor Laboratory Press: Cold Spring Harbor, NY, USA, 1989.
39. Bradford, M.M. A rapid and sensitive method for the quantitation of microgram quantities of protein utilizing the principle of protein-dye binding. *Anal. Biochem.* **1976**, *72*, 248–254. [CrossRef]
40. Laemmli, U.K. Cleavage of structural proteins during the assembly of the head of bacteriophage T4. *Nature* **1970**, *227*, 680–685. [CrossRef]

International Journal of
Molecular Sciences

MDPI

Article

2-Ketoglutarate-Generated In Vitro Enzymatic Biosystem Facilitates Fe(II)/2-Ketoglutarate-Dependent Dioxygenase-Mediated C–H Bond Oxidation for (2s,3r,4s)-4-Hydroxyisoleucine Synthesis

Xiao-Ran Jing [1], Huan Liu [1], Yao Nie [1,*] and Yan Xu [1,2,*]

1 Key Laboratory of Industrial Biotechnology of Ministry of Education and School of Biotechnology, Jiangnan University, 1800 Lihu Road, Wuxi 214122, China; rushjxr794@163.com (X.-R.J.); 18635924851@163.com (H.L.)

2 State Key Laboratory of Food Science and Technology, Jiangnan University, 1800 Lihu Road, Wuxi 214122, China

* Correspondence: ynie@jiangnan.edu.cn (Y.N.); yxu@jiangnan.edu.cn (Y.X.); Tel.: +86-510-85197760 (Y.N.); +86-510-85918201 (Y.X.)

Received: 24 June 2020; Accepted: 20 July 2020; Published: 28 July 2020

Abstract: Fe(II)/2-ketoglutarate-dependent dioxygenase (Fe(II)/2-KG DO)-mediated hydroxylation is a critical type of C–H bond functionalization for synthesizing hydroxy amino acids used as pharmaceutical raw materials and precursors. However, DO activity requires 2-ketoglutarate (2-KG), lack of which reduces the efficiency of Fe(II)/2-KG DO-mediated hydroxylation. Here, we conducted multi-enzymatic syntheses of hydroxy amino acids. Using (2s,3r,4s)-4-hydroxyisoleucine (4-HIL) as a model product, we coupled regio- and stereo-selective hydroxylation of L-Ile by the dioxygenase IDO with 2-KG generation from readily available L-Glu by L-glutamate oxidase (LGOX) and catalase (CAT). In the one-pot system, H_2O_2 significantly inhibited IDO activity and elevated Fe^{2+} concentrations of severely repressed LGOX. A sequential cascade reaction was preferable to a single-step process as CAT in the former system hydrolyzed H_2O_2. We obtained 465 mM 4-HIL at 93% yield in the two-step system. Moreover, this process facilitated C–H hydroxylation of several hydrophobic aliphatic amino acids to produce hydroxy amino acids, and C–H sulfoxidation of sulfur-containing L-amino acids to yield L-amino acid sulfoxides. Thus, we constructed an efficient cascade reaction to produce 4-HIL by providing prerequisite 2-KG from cheap and plentiful L-Glu and developed a strategy for creating enzymatic systems catalyzing 2-KG-dependent reactions in sustainable bioprocesses that synthesize other functional compounds.

Keywords: Fe(II)/2-ketoglutarate-dependent dioxygenase; 2-ketoglutarate generation; regio- and stereo-selective synthesis; hydroxy amino acids; sequential cascade reaction

1. Introduction

The C–H functionalization of small molecules is an important reaction in organic synthesis [1,2]. The oxidized products were used as building blocks in pharmaceutical syntheses [3]. The C–H bond activation by biocatalysis has been reported to date, with less environmental impact compared with conventional counterparts [4]. Fe (II)- and 2-ketoglutarate-dependent dioxygenases (Fe(II)/2-KG DOs) catalyze various C–H-activation-mediated reactions including hydroxylation, deoxygenation, desaturation, and ring extension [5]. Recently, research attention has been directed to the asymmetric hydroxylation of inactivated carbon atoms via direct C–H bond functionalization [1].

Amino acids are substrates of Fe(II)/2-KG DOs and are usually converted into hydroxy amino acids [6]. Various amino acid hydroxylases producing hydroxy amino acids have been characterized among the Fe(II)/2-KG DOs. These include lysine hydroxylase [7], asparagine hydroxylase [8], L-Ile dioxygenase (IDO) [9], leucine hydroxylase [10], and proline hydroxylase [11,12].

Hydroxy amino acids have numerous applications in the pharmaceutical and food industries. They serve as chiral building blocks and food additives [13]. For example, *trans*-4-hydroxy-L-proline (*trans*-4-Hyp) is a major component of collagen [14]. It is a crucial precursor in the pharmaceutical synthesis of N-aryl pyrrole [15] and (−)-kainic acid [16]. Moreover, (2S,3R,4S)-4-hydroxyisoleucine (4-HIL) naturally occurs in *Trigonella foenum-graecum* (fenugreek) seeds, accelerates insulin secretion, and could be administered to treat type II diabetes [17]. In addition, 4-hydroxynorvaline isolated from *Lathyrus odoratus* seeds [18] stimulates insulin secretion and has antidiabetic properties [19]. Further, 5-hydroxyleucine participates in griselimycin synthesis. Griselimycin may have efficacy against drug-resistant *Mycobacterium tuberculosis* and strong antituberculosis activity [10].

Fe(II)/2-KG-DOs catalyze C–H oxidation using 2-KG as a co-substrate. Thus, the amount of 2-KG directly influences the catalytic efficiency of Fe(II)/2-KG DOs. Several studies applied metabolic engineering to generate 2-KG and facilitate Fe(II)/2-KG DO-mediated C–H bond oxidation. Smirnov et al. redirected the tricarboxylic acid (TCA) cycle intermediate 2-KG to biotransform L-Ile into 4-HIL using IDO. They achieved 82% yield using 100 mM L-Ile as the substrate [20]. Zhang et al. optimized the TCA cycle by dynamically modulating the activity of the 2-ketoglutarate dehydrogenase complex in *Corynebacterium glutamicum*. In this manner, they generated ample 2-KG and yielded 232.52 mM of 4-HIL after 64 h [21]. A similar strategy was applied to other Fe(II)/2-KG-DOs-mediated reactions in order to overcome insufficient 2-KG supply. Lin et al. reported a deacetoxycephalosporin-C synthase (DAOCS) that converted penicillin substrates into cephalosporins. They reconstituted the TCA cycle to force 2-KG into an enzymatic reaction [22]. However, this modification had a negative effect on cell growth [21,22], low conversion efficiency, and long reaction time. In contrast, 2-KG has also been produced by chemical synthesis and enzymatic oxidation [23,24]. Enzymatic 2-KG syntheses were conducted under relatively mild reaction conditions, consumed minimal energy, and produced little pollution. Liu et al. reported the use of L-amino acid oxidase (LAAO) for 2-KG production with L-glutamic acid (L-Glu) as the substrate. Nevertheless, LAAO had low activity towards L-Glu and was inhibited by high substrate and product concentrations. L-glutamate oxidase (LGOX) has been identified with high L-Glu-to-2-KG conversion activity without requiring exogenous (flavin adenine dinucleotide) FAD [25]. Since the oxidative deamination from L-Glu to 2-KG is along with H_2O_2 generation, Wu et al. developed a cascade strategy of co-expression LGOX and catalase (CAT) to eliminate H_2O_2 [26].

In vitro and in vivo (whole-cell) enzymatic syntheses of complex molecules have been successfully implemented in recent years [27]. Here, we implemented an in vitro three-enzyme system to synthesize hydroxy amino acids. We used IDO to hydroxylate L-amino acids and performed oxidative decarboxylation of 2-KG to succinate (SA) [28]. As the consumption of superstoichiometric amount of 2-KG is one of the limitations of the system, an adequate supply of 2-KG derived from inexpensive substrates is necessary for an in vitro system. Hence, 2-KG generation from L-Glu catalyzed by LGOX was adopted in combination with CAT to eliminate the H_2O_2 by-product. An in vitro multi-enzymatic system was successfully established to synthesize the novel insulin secretion accelerant 4-HIL for the treatment of type II diabetes. Moreover, this system is highly efficient at catalyzing reactions involving other hydrophobic aliphatic L-amino acids and could synthesize hydroxy amino acids [28]. Here, Fe(II)/2-KG DOs-mediated C–H oxidation of small molecules was accomplished in vitro using an efficient and sustainable 2-KG supply system that could generate this substrate for other enzymes catalyzing the synthesis of various functional compounds.

2. Results and Discussion

2.1. Design of a Fe(II)/2-KG DO-Based In Vitro Enzymatic Biosystem for Hydroxy Amino Acid Synthesis

Fe(II)/2-KG DOs catalyze various reactions including hydroxylation by C–H activation. Binding of the 2-KG co-substrate facilitates substrate binding and forms a ferryl intermediate that is critical for substrate hydroxylation [29]. The quantity of 2-KG directly affects the catalytic efficiency of Fe(II)/2-KG DOs. Thus, we implemented an in vitro bio-cascade reaction system to synthesize hydroxy amino acids by coupling Fe(II)/2-KG DOs-mediated C–H hydroxylation with 2-KG derived from cheap, readily available L-Glu (Figure 1). In this multi-enzymatic cascade system, IDO, the Fe(II)/2-KG DO, catalyzes L-amino acid hydroxylation and synthesizes useful hydroxy amino acids and certain sulfur-containing L-amino acids. This process is accompanied by oxidative 2-KG decarboxylation using Fe^{2+} as a cofactor [28]. LGOX exhibits high specificity for the conversion of L-Glu to 2-KG and simultaneously generates H_2O_2. Catalase (CAT) eliminates the H_2O_2 [25], thereby preventing its potentially negative effects on the cascade reaction such as the oxidization of 2-KG to succinic acid. Overall, hydroxy amino acid synthesis is accomplished with a simultaneous 2-KG supply by constructing a cell-free reaction system with no additional substrate consumption or membrane-induced mass transfer [30]. Compared with in vivo 2-KG accumulation from the TCA cycle, in vitro cell-free enzymatic system for 2-KG generation has the advantages of high space-time yield and easy operation in product separation. Moreover, this strategy is applicable to other enzymes that catalyze 2-KG-coupled reactions and synthesize other functional compounds.

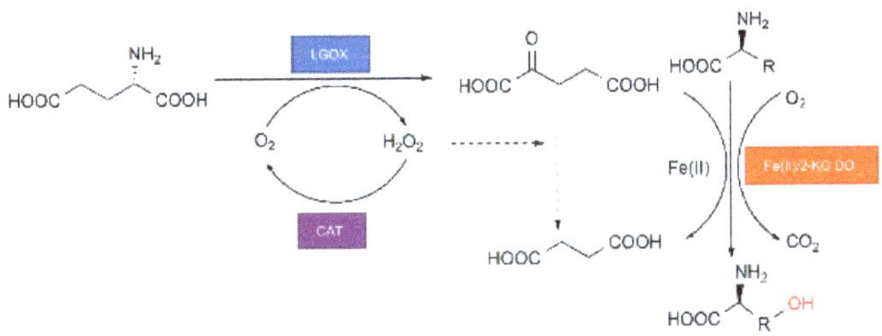

Figure 1. Multi-enzymatic cascade system for hydroxy amino acid synthesis by combining Fe(II)/2-ketoglutarate-dependent dioxygenase (Fe(II)/2-KG DO), L-glutamate oxidase (LGOX), and catalase (CAT).

2.2. Enzyme Kinetics

IDO and LGOX in the cascade reaction were purified by His-Trap HP affinity chromatography (Figure S1), and we optimized the reaction conditions to achieve maximum system efficiency (Figure S2). According to the optimized results, we observed the kinetics of IDO and LGOX under 50 mM Tris HCl, pH 7.0, and 30 °C. In these conditions, the IDO and LGOX activity levels were relatively high. The kinetic parameters were fitted to a Michaelis–Menten model (Table 1). IDO exhibited high oxidative decarboxylation activity on 2-KG (V_{max} = 19.79 ± 0.08 U·mg^{-1}). It also had relatively high activity in the presence of L-Ile (V_{max} = 8.99 ± 0.09 U·mg^{-1}). In contrast, L-Ile hydroxylase from *Bacillus cereus* 13658 had an activity of only 0.68 ± 0.06 U·mg^{-1} [31]. LGOX had high specificity and affinity for L-Glu (K_m = 2.65 ± 0.11 mM and V_{max} = 2.93 U·mg^{-1}). IDO and LGOX did not markedly differ in terms of k_{cat}. Therefore, the rate of 2-KG regeneration aligned with that of L-Ile hydroxylation and facilitated 4-HIL synthesis.

Table 1. Kinetic parameters of IDO and LGOX.

Enzyme	Substrate	K_m (mM)	V_{max} (U·mg^{-1})	k_{cat} (s^{-1})	k_{cat}/K_m (s^{-1}·m·M^{-1})
IDO	L-Ile	6.34 ± 0.12	8.99 ± 0.09	4.18 ± 0.08	0.66 ± 0.04
	2-KG	15.12 ± 0.08	19.79 ± 0.08	9.18 ± 0.07	0.61 ± 0.07
LGOX	L-Glu	2.65 ± 0.11	2.93 ± 0.06	3.38 ± 0.05	1.28 ± 0.07

Kinetic parameters were measured in 50 mM Tris-HCl (pH 7.0) at 30 °C.

2.3. *Effects of Reaction Components on IDO and LGOX Activity*

There are generally some factors to be considered for constructing a cascade reaction involving different reaction components [32]. Various intermediates and components may influence the reaction process [33]. We investigated the inhibitory effects of L-Glu, L-Ile, SA, L-ascorbic acid (Vc), Fe^{2+} ($FeSO_4{\cdot}7H_2O$), and H_2O_2 on single enzyme activity by varying their concentrations. Figure 2 shows that H_2O_2 strongly inhibited IDO and significantly decreased its activity. Exposure of the enzyme to the oxidizing environment created by H_2O_2 oxidized the Fe^{2+} in the IDO active site and decreased enzyme activity. Vc is a reductant that promotes C–H oxidation catalyzed by Fe(II)/2-KG DOs [34]. Vc was required by this system for 4-HIL synthesis. Further, 50 mM SA lowered IDO activity to 74%. However, all other components tested only slightly affected IDO activity. High Fe^{2+} concentrations inactivate LGOX [25]. The relative activity of LGOX was <20% at 20 mM Fe^{2+}. IDO is a member of the Fe (II)/2-KG-dependent dioxygenase family. It requires 2-KG and Fe^{2+} as cofactors to catalyze hydroxylation. Here, IDO exhibited little activity towards the substrate even in the absence of Fe^{2+}, possibly because of the Fe^{2+} pool in the *E. coli* cells [35]. Therefore, the influence of Fe^{2+} addition on the cascade reaction needs to be further investigated.

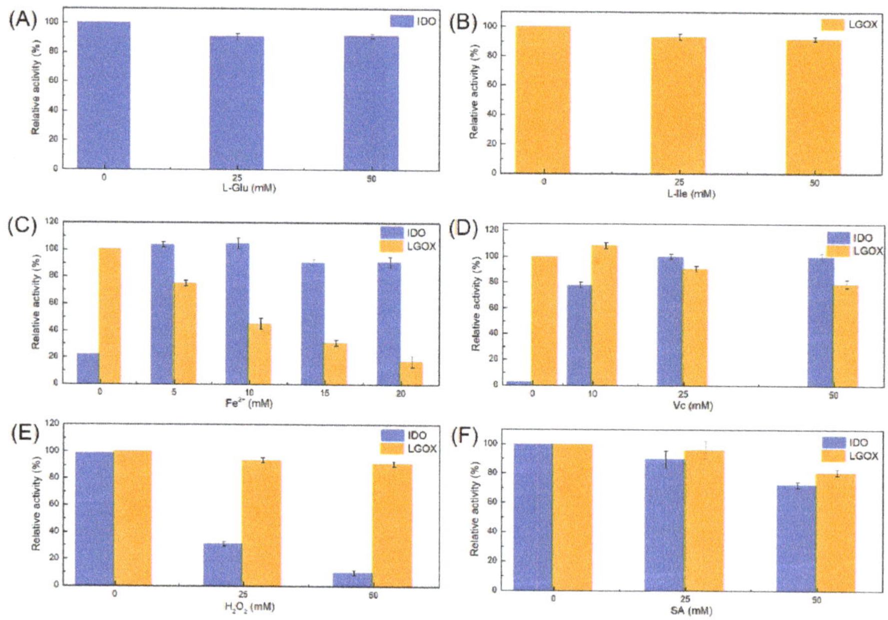

Figure 2. Effects of various reaction components on IDO (blue) and LGOX (orange) activity. IDO and LGOX activity levels were measured using various concentrations of (**A**) L-Glu, (**B**) L-Ile, (**C**) Fe^{2+}, (**D**) Vc, (**E**) H_2O_2, and (**F**) SA. The specific activities of the enzymes under standard conditions (5.84 ± 0.07 U·mg^{-1} for IDO and 2.67 ± 0.09 U·mg^{-1} for LGOX) were designated as 100%, respectively.

2.4. Enzymatic Cascade for 4-HIL Synthesis

Here, we attempted to develop an in vitro system producing hydroxy amino acids and using IDO, LGOX, and CAT as biocatalysts. We tested one-pot 4-HIL production using a combination of the aforementioned enzymes (Figure 3A).

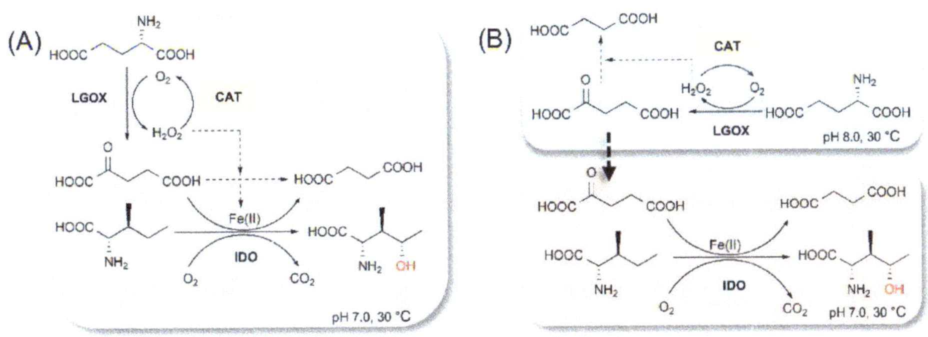

Figure 3. Enzymatic cascade production of 4-HIL. (**A**) Scheme of one-pot production of 4-HIL. (**B**) Scheme of multistep reaction to produce 4-HIL.

We assessed the influence of exogenous Fe^{2+} on 4-HIL production. We mixed 100 mM L-Glu and 100 mM L-Ile in a 25 mL shaker flask and added 1 g·L^{-1} IDO, 0.5 g·L^{-1} LGOX, and 2 mg·L^{-1} CAT to it. Figure 4 shows that 166 mM of 4-HIL was produced in the absence of Fe^{2+} whereas 28 mM of 4-HIL was produced in the presence of Fe^{2+}. IDO and LGOX activity obviously decreased in the cascade reaction relative to the single-step reaction (control). The final 4-HIL product did not continue to accumulate after the first hour of the reaction. The reaction mixture presented with low 4-HIL yield (16%) in the absence of Fe^{2+} and low IDO activity (Figure 4A). As shown in Figure 4B, Fe^{2+} obviously inhibited LGOX activity. Minimal L-Glu was consumed and little 4-HIL accumulated after 0.5 h. However, residual 2-KG was detected in the reaction mixture after sufficient CAT was added to eliminate the H_2O_2. IDO and LGOX were simultaneously inhibited by H_2O_2 and Fe^{2+}, respectively. Therefore, the generation of 2-KG from L-Glu by LGOX was incompatible with the hydroxylation of L-Ile by IDO in one pot. For this reason, it was preferable to conduct a stepwise cascade reaction to produce hydroxy amino acids efficiently.

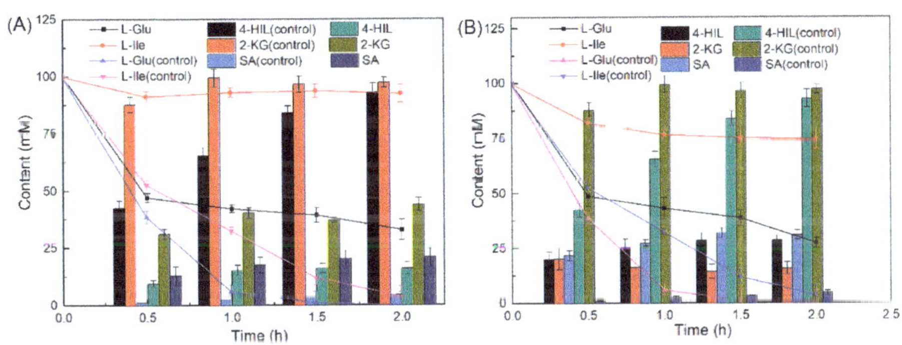

Figure 4. One-pot cascade reaction to produce 4-HIL. (**A**) One-pot cascade reaction in the absence of Fe^{2+}; (**B**) one-pot cascade reaction in the presence of Fe^{2+}.

We then attempted two-step 4-HIL production (Figure 3B) and further optimized the biocatalyst and substrate concentrations. L-Ile in the concentration range of 100–500 mM was hydroxylated with

1 g·L^{-1} purified IDO. Figure 5A shows that IDO had robust L-Ile hydroxylation activity. We obtained 444 mM of 4-HIL with 500 mM L-Ile as substrate. Thus, the product yield was 89%. Moreover, no substrate inhibition was detected at any substrate concentration. We examined the effects of IDO on L-Ile transformation at various substrate concentrations. The corresponding time courses are shown in Figure 5B. The yield and conversion rate increased with IDO concentration. After 9 h, 2 g·L^{-1} IDO generated 461 mM of 4-HIL from 500 mM L-Ile at a product yield of 92%. After 9 h, 1 g·L^{-1} IDO generated 459 mM of 4-HIL at a product yield of 92%. However, concerning the reaction productivity, 4.9 g·(L·g·h)$^{-1}$ of 4-HIL was obtained with 2 g·L^{-1} IDO, while 7.5 g·(L·g·h)$^{-1}$ of 4-HIL was obtained with 1 g·L^{-1} IDO. Therefore, 1 g·L^{-1} IDO was selected as the catalyst for L-Ile hydroxylation. Hydrogen peroxide is co-produced with 2-KG and may degrade it to SA. The H_2O_2 is decomposed to H_2O and O_2 by CAT [36]. We explored the effect of varying CAT concentration on 2-KG production using 200 mM L-Glu as the substrate. Figure 5C shows that the SA accumulation was <15 mM in the presence of 2–3 mg·L^{-1} CAT. The 2-KG production slightly increased with increasing CAT concentration. At 2 mg·L^{-1} CAT, 186 mM of 2-KG was generated within 120 min and the yield was 93%. In contrast, only 2.5% SA was formed. No further increase in 2-KG was observed with increasing CAT concentration. We also tested various L-Glu concentrations to establish the optimal reaction conditions (Figure 5D). There was slight substrate inhibition with increasing L-Glu concentration. LGOX generated 491 mM of 2-KG in the presence of 550 mM L-Glu and the yield was 90%.

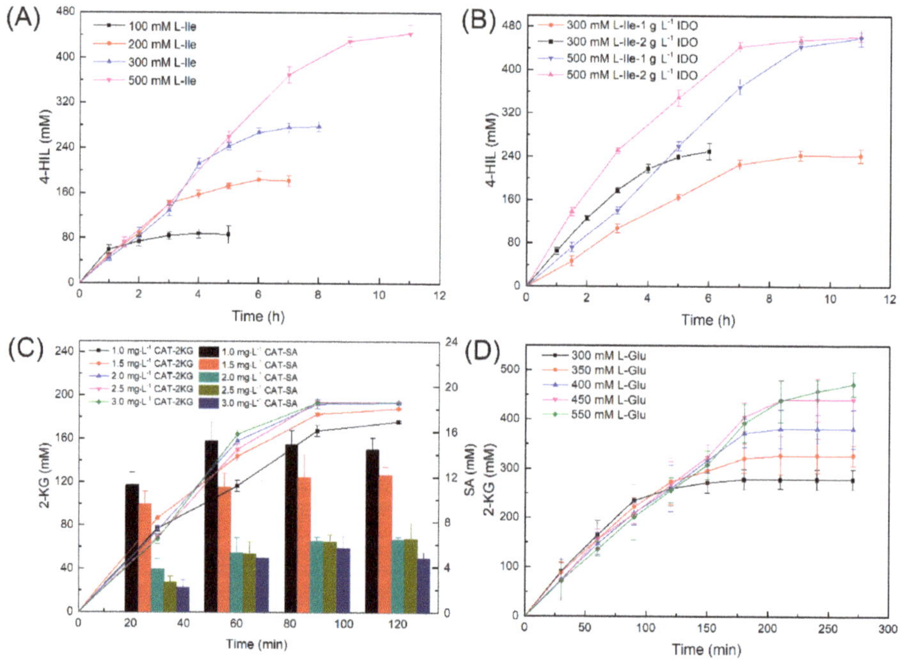

Figure 5. Optimization of biocatalyst and substrate concentrations. (**A**) 4-HIL production at different L-Ile concentrations with 1 g·L^{-1} IDO; (**B**) 4-HIL production at various IDO concentrations; (**C**) 2-KG and SA yield at different CAT concentrations in presence of 0.5 g·L^{-1} LGOX; (**D**) 2-KG yield at various L-Glu concentrations in presence of 0.5 g·L^{-1} LGOX.

2.5. Fed-Batch 4-HIL Synthesis Reaction

L-Ile is the most hydrophobic amino acid. Its solubility is 32.5 g·L^{-1} (247.75 mM) in aqueous solution at 25 °C [37]. To further improve transformation efficiency of cell-free enzymatic system for

4-HIL synthesis, we explored methods of feeding L-Ile in a 500 mL bioreactor (Figure 6). L-Glu at 550 mM was fully converted to 2-KG in ~5 h and only <35 mM SA accumulated. In the first oxidation reaction, 491 mM of 2-KG was obtained at a reaction rate of 14.5 $g \cdot L^{-1} \cdot h^{-1}$. Optimal concentrations of the L-Ile hydroxylation components were added to the system and 500 mM L-Ile was added either in one dose or in batches at an initial concentration of 200 mM. IDO maintained robust L-Ile hydroxylation activity when the substrate was added in one dose. We obtained 434 mM of 4-HIL after 6 h of L-Ile hydroxylation. The L-Ile hydroxylation reaction rate significantly decreased when the initial substrate concentration was 200 mM as the L-Ile was rapidly consumed. Nevertheless, the reaction rate increased after substrate supplementation. In the second hydroxylation reaction, when L-Ile was incrementally supplemented, IDO yielded 440 mM of 4-HIL after 9 h. In contrast, IDO produced 465 mM of 4-HIL after the L-Ile was added in one dose. The total reaction time for the entire cascade reaction was 14.5 h. The first step produced 491 mM of 2-KG from L-Glu within 5 h and the yield was 89%. The second step generated 464.96 mM of 4-HIL after 9 h when L-Ile was added in one dose. The yield was 93%. Hence, it is feasible to scale up the foregoing 4-HIL production system. In contrast, an in vivo 2-KG supply for 4-HIL synthesis has been successfully implemented by redirecting the TCA cycle [21,38]. However, metabolic engineering strategies of shifting the carbon flux from L-Ile to 2-KG by boosting the TCA cycle and further increasing the pool of oxaloacetate were adopted to obtain 34.21 g/L (232.45 mM) of 4-HIL at 64 h, inevitably hindering normal cell growth [21]. The cell-free enzymatic system generating 2-KG from L-Glu was favorable for 4-HIL production with high substrate concentration and space–time yield. Furthermore, the methodology optimized here could lay the foundation for the development of other industrial bioprocess technologies.

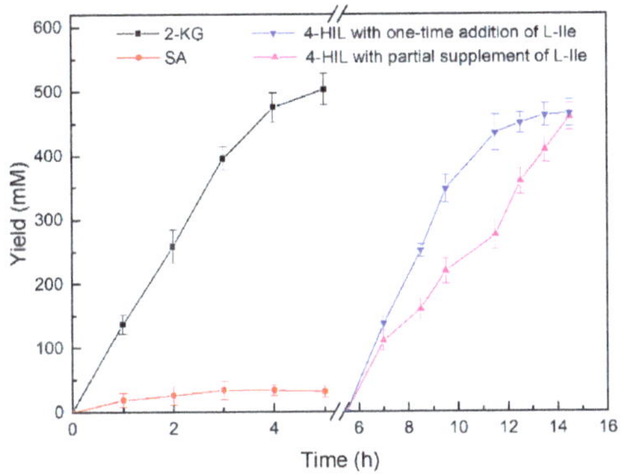

Figure 6. Two-step reaction for 4-HIL production.

2.6. In Vitro Enzymatic Biosystem for Hydroxy Amino Acids Synthesis

Hydroxy amino acids are active ingredients or building blocks in pharmaceutical synthesis. Thus, effective, sustainable hydroxy amino acid production is urgently required. IDO has enzymatic activity towards certain hydrophobic aliphatic L-amino acids and sulfur-containing amino acids [28]. It hydroxylates or sulfonates C–H bonds and generates hydroxy amino acids and L-amino acids sulfoxides, respectively. Here, we used 100 mM of L-amino acids substrates in the presence of IDO. All enzymatic reactions achieved high substrate conversion rates. Figure 7 shows that LGOX generated 102.19 mM of 2-KG from L-Glu within 1 h. IDO produced 16.47 $g \cdot L^{-1}$ of L-methionine sulfoxide (>99.9% conversion) within the first 0.5 h in the presence of L-methionine (L-Met). IDO generated 14.61 $g \cdot L^{-1}$ of 4-hydroxy-L-leucine (>99.9% conversion) within 2 h in the presence of L-leucine (L-Leu). IDO converted 89.2% of L-norvaline (L-Nva) after 3 h. When L-norleucine (L-Nle) was the substrate, the putative diastereomers 4-hydroxy-L-norleucine and 5-hydroxy-L-norleucine were detected. The reaction products were characterized by mass spectrometry (Figure S3). The aforementioned reactions demonstrated that the application of a two-step cascade system was highly efficient for the production of hydroxy amino acids and L-amino acid sulfoxides via direct C–H functionalization.

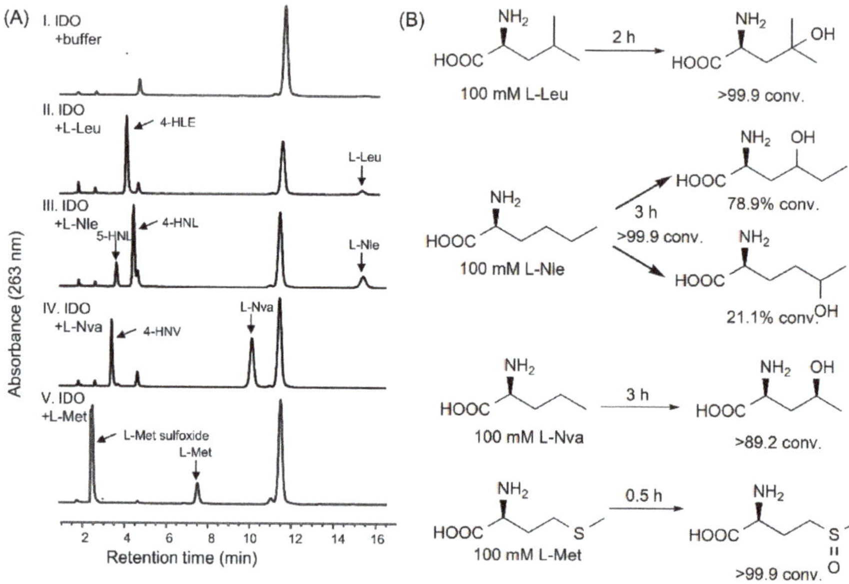

Figure 7. In vitro enzymatic synthesis of hydroxy amino acids. (**A**) Hydroxy amino acids were generated from a two-step system using various L-amino acids as substrates. Products were analyzed by LC-MS. (**B**) Biotransformation data for the productions of different hydroxy amino acids.

3. Materials and Methods

3.1. Reagents

The (2s,3R,4s)-4-hydroxyisoleucine standard was purchased from Cambridge Sigma-Aldrich (Munich, Germany). All other analytical-grade chemicals were obtained from Sinopharm Chemical Reagent Co. (Shanghai, China). All enzymes for genetic manipulations were obtained from TaKaRa Biotechnology Co. (Dalian, China). Protein purification columns were acquired from GE Healthcare (Munich, Germany).

3.2. Preparation of Recombinant Enzymes

Commercial catalase (CAT) was obtained from Aladdin Bio-Chem Technology Co. Ltd. (Shanghai, China). The 6×His-tagged IDO-expressing strain (*E. coli* BL21(DE3)-IDO) [28] and 6×His-tagged LGOX-expressing strain (*E. coli* BL21(DE3)-LGOX) [25,26] were constructed as previously described. IDO and LGOX precultures were incubated overnight in 5 mL of Luria–Bertani (LB) medium (Sangon Biotech, Shanghai, China) with 50 μg·mL^{-1} kanamycin (Sinopharm Chemical Reagent, Shanghai, China) at 37 °C on a rotary shaker (200 rpm). IDO and LGOX were induced with 0.5 mM isopropyl-β-D-1-thiogalactopyranoside (IPTG) in 200 mL of LB medium containing 50 μg·mL^{-1} kanamycin. The cultures were incubated for 3 h at 37 °C and then for 16 h at 17 °C.

The cells were harvested by centrifugation (10,800× *g*, 4 °C, 5 min) and resuspended in 20 mM Tris-HCl buffer (pH 7.5). The bacterial pellet was disrupted and centrifuged (10,800× *g*, 4 °C, 40 min). The recombinant IDO and LGOX supernatants were purified by Ni-affinity chromatography as previously reported [39]. The supernatant was purified in a His-Trap HP affinity column (GE Healthcare, Little Chalfont, UK) [40]. The elution buffer in the affinity columns created a high salt content in the purified fractions. The excess salt was removed with disposable PD-10 desalting columns (GE Healthcare, Little Chalfont, UK) [41]. The proteins were analyzed by sodium dodecyl sulfate-polyacrylamide gel electrophoresis (SDS-PAGE) [42] and the yields were determined by NanoDrop 8000 Microvolume UV-Vis spectrophotometry (Thermo Fisher Scientific, Waltham, MA, USA).

3.3. Enzymatic Activity Assays

For IDO, a mixture of 10 mM L-Ile, 1.5 mM $FeSO_4$, 10 mM L-ascorbic acid, 10 mM 2-KG, and 50 mM Tris-HCl buffer (pH 7.0) was incubated in a Thermomixer Comfort incubator (Eppendorf; Hamburg, Germany) at 30 °C with 800 rpm agitation. The reaction was initiated with 0.1 $g{\cdot}L^{-1}$ purified IDO in a final volume of 300 μL [43]. Samples were collected at 0, 5, and 10 min after enzyme addition and analyzed by HPLC [44]. One unit of enzymatic activity was defined as the amount of enzyme required to catalyze the synthesis of 1 μmol of 4-HIL min^{-1} under standard conditions.

For LGOX, a mixture of 10 mM L-Glu, 1 $mg{\cdot}L^{-1}$ CAT, and 50 mM Tris-HCl buffer (pH 7.0) was incubated in a Thermomixer Comfort incubator (Eppendorf; Hamburg, Germany) at 30 °C, with 800 rpm agitation. The reaction was initiated with 0.1 $g{\cdot}L^{-1}$ purified LGOX in a final volume of 300 μL. Samples were collected at 1, 3, and 5 min after enzyme addition and analyzed by HPLC. One unit of enzymatic activity was defined as the amount of enzyme required to catalyze the synthesis of 1 μmol of 2-KG· min^{-1} under standard conditions.

3.4. Determination of K_m and V_{max}

For IDO, the optimal concentration of the cascade reaction was used and the L-Ile concentration range was 0.5–20 mM. The concentration of 2-KG was in the range of 1–40 mM. The kinetic parameter assays were conducted in triplicate. For LGOX, the optimal concentration of the cascade reaction was used and the L-Glu concentration range was 0.5–20 mM. IDO and LGOX activity were measured as previously described. The kinetic parameters were fitted to a Michaelis–Menten model [45].

3.5. Effects of Reaction Components on Enzyme Activities

The inhibitory effects of L-Ile, L-Glu, Fe^{2+}, Vc, H_2O_2, and SA were tested by comparing the initial reaction rates of IDO and LGOX in the presence of various concentration of the aforementioned compounds. IDO and LGOX activity were measured as previously described.

3.6. One-Pot Synthesis of 4-HIL

The reaction mixture consisted of 100 mM L-Ile, 50 mM Vc, 100 mM L-Glu, 1 $g{\cdot}L^{-1}$ IDO, 0.5 $g{\cdot}L^{-1}$ LGOX, 2 $mg{\cdot}L^{-1}$ CAT, and 50 mM Tris-HCl buffer (pH 7.0). Then 5 mM $FeSO_4{\cdot}7H_2O$ was added to the mixture to compare the effects of Fe^{2+} on the cascade reaction. The system comprised 5 mL of the reaction mixture heated to 30 °C in a 25-mL shaker flask rotating at 200 rpm.

3.7. Two-Step Process for Hydroxy Amino Acid Synthesis

For the 2-KG regeneration step, the reaction mixture consisted of 0.5 $g{\cdot}L^{-1}$ LGOX, various concentrations of L-Glu (300–550 mM) and CAT (1-3 $mg{\cdot}L^{-1}$), and 50 mM Tris-HCl buffer (pH 8.0) in a final volume of 5 mL. The reaction was conducted at 30 °C and 200 rpm. For the L-amino acid hydroxylation step, L-amino acids hydroxylation by IDO was examined after 2-KG production by LGOX. The reaction mixture comprised 5 mM $FeSO_4$, 50 mM Vc, 1 $g{\cdot}L^{-1}$ IDO, and 50 mM Tris-HCl buffer (pH 7.0) in a final volume of 5 mL. Using various substrate concentrations (100–500 mM), the 2-KG obtained from the first step was added in equimolar concentrations with the L-amino acids. The mixture was incubated in a shaker flask at 30 °C and 200 rpm.

3.8. Fed-Batch Reaction for 4-HIL Synthesis

For 2-KG generation, the reaction mixture was added to a 500-mL reactor consisting of 0.5 g·L^{-1} LGOX, 550 mM L-Glu, 2 mg·L^{-1} CAT, and 50 mM Tris-HCl buffer (pH 8.0) in a final volume of 100 mL. The reaction was conducted at 30 °C and 400 rpm. After 5 h, the components containing 10 mM $FeSO_4$, 50 mM Vc, and 1 g·L^{-1} IDO were mixed into the reactor. The pH was fixed at 7.0. The one-dose and incremental L-Ile addition methods were further compared as L-Ile has low solubility in aqueous solvents. L-Ile 500 mM was added to the reactor in one dose to synthesize 4-HIL. The fed-batch reaction was conducted using 200 mM L-Ile as the initial concentration. Then 150 mM L-Ile was supplemented every 2 h until a final concentration of 500 mM L-Ile was reached. The reactions were conducted at 30 °C and 400 rpm. The supernatants were sampled every hour and analyzed.

3.9. Analysis of Organic Acids

The organic acids in the samples were analyzed in a Waters 2695 HPLC system (Waters Co., Milford, MA, USA) fitted with an Atlantis T3 column (4.6 mm × 250 mm). Compounds were detected at 210 nm using an injection volume of 10 μL and a column temperature of 40 °C. The mobile phase was 20 mM phosphate buffer (pH 2.8) and the flow rate was 0.8 mL·min^{-1}.

3.10. Analysis of Amino Acids

The samples were analyzed in a Waters 2695 HPLC system (Waters Co., Milford, MA, USA) fitted with a Diomansil C18 column (4.6 mm × 250 mm). The chromatographic conditions were as follows: mobile phase A (NaAc-HAc buffer (50 mM, pH 4.2):acetonitrile = 50:50); mobile phase B (acetonitrile); gradient elution program; flow rate, 1 mL min^{-1}; column temperature, 25 °C; injection volume, 10 μL. Post-column derivatization was conducted with Fmoc-Cl [44]. The Fmoc-Cl derivatives of the amino acids were detected at 263 nm.

LC-MS analysis was conducted in a Waters ACQUITY UPLC-MS system fitted with a Waters ACQUITY UPLC HSS C18 reverse-phase column (inner diameter, 1.8 μm) (Waters Co., Milford, MA, USA). The inlet, MS transfer line, and ion source temperatures were set to 280, 280, and 230 °C, respectively.

3.11. Nuclear Magnetic Resonance (NMR) Analysis

All products were isolated by cation exchange chromatography using a strong cation resin (C100E (H^+ form), Purolite, King of Prussia, PA, USA) [46]. L-Ile hydroxylation generated 40 mM of 4-HIL and 42% yield. The 4-HIL spectrum was recorded by ^{1}H-NMR spectroscopy (Figure S4) as previously described [9].

(2S,3R,4S)-4-hydroxyisoleucine: δ 3.84 (d, J = 4.4 Hz, 1H), 3.79 (dd, J = 13.6, 6.8 Hz, 1H), 1.92–1.82 (m, 1H), 1.19 (d, J = 6.3 Hz, 3H), 0.91 (d, J = 7.1 Hz, 3H).

4. Conclusions

Here, we developed an in vitro enzymatic system to synthesize hydroxy amino acids via Fe(II)/2-KG DOs-mediated C–H bond oxidation. Amino acid hydroxylation by Fe(II)/2-KG DO was coupled with generation of the prerequisite 2-KG from inexpensive and abundant L-Glu. After systematically assessing the catalytic parameters of the catalysts, a sequential cascade reaction was preferable for the efficient synthesis of hydroxy amino acids. Optimizing the catalyst and substrate concentrations in the two-step process resulted in 465 mM of 4-HIL and 93% yield. Otherwise, IDO was highly active towards other hydrophobic aliphatic L-amino acids and could, therefore, produce other hydroxy amino acids from them. Hence, this system is promising for in vitro biosynthesis of hydroxy amino acids, and more efforts should be necessary to enhance the resistance of IDO and LGOX towards the inhibition of reaction generated components, so as to facilitate the one-pot reaction system. It is also a viable alternative to other Fe(II)/2-KG DO-mediated C–H bond functionalization strategies for the synthesis of functional compounds.

Supplementary Materials: Supplementary materials can be found at http://www.mdpi.com/1422-0067/21/15/5347/s1.

Author Contributions: Conceptualization, X.-R.J. and Y.N.; methodology, X.-R.J.; software, X.-R.J.; validation, X.-R.J. and H.L.; formal analysis, X.-R.J.; investigation, X.-R.J.; resources, Y.N. and Y.X.; data curation, X.-R.J. and H.L.; writing—original draft preparation, X.-R.J., Y.N. and Y.X.; writing—review and editing, X.-R.J., Y.N. and Y.X.; visualization, X.-R.J., Y.N. and Y.X.; supervision, Y.N. and Y.X.; project administration, Y.N. and Y.X.; funding acquisition, Y.N. and Y.X. All authors have read and agreed to the published version of the manuscript.

Funding: This work was financially supported by the National Natural Science Foundation of China (NSFC) (No. 21676120), the 111 Project (No. 111-2-06), the High-End Foreign Experts Recruitment Program (No. G20190010083), the Program for Advanced Talents within Six Industries of Jiangsu Province (No. 2015-NY-007), the National Program for Support of Top-Notch Young Professionals, the Fundamental Research Funds for the Central Universities (No. JUSRP51504), the Project Funded by the Priority Academic Program Development of Jiangsu Higher Education Institutions, Top-Notch Academic Programs Project of Jiangsu Higher Education Institutions, the Jiangsu Province "Collaborative Innovation Center for Advanced Industrial Fermentation" Industry Development Program, the Postgraduate Research & Practice Innovation Program of Jiangsu Province (No. KYCX19_1833), the Program for the Key Laboratory of Enzymes of Suqian (No. M201803), and the National First-Class Discipline Program of Light Industry Technology and Engineering (No. LITE2018-09).

Conflicts of Interest: There are no conflicts of interest to declare.

References

1. Newhouse, T.; Baran, P.S. If C-H bonds could talk: Selective C-H Bond oxidation. *Angew. Chem. Int. Ed. Engl.* **2011**, *50*, 3362–3374. [CrossRef] [PubMed]
2. Davies, H.M.L.; Morton, D. Recent advances in C-H functionalization. *J. Org. Chem.* **2016**, *81*, 343–350. [CrossRef] [PubMed]
3. Qiu, Y.; Gao, S. Trends in applying C-H oxidation to the total synthesis of natural products. *Nat. Prod. Rep.* **2016**, *33*, 562–581. [CrossRef] [PubMed]
4. Dong, J.; Fernandez-Fueyo, E.; Hollmann, F.; Paul, C.E.; Pesic, M.; Schmidt, S.; Wang, Y.; Younes, S.; Zhang, W. Biocatalytic oxidation reactions: A chemist's perspective. *Angew. Chem. Int. Ed. Engl.* **2018**, *57*, 9238–9261. [CrossRef] [PubMed]
5. Martinez, S.; Hausinger, R.P. Catalytic mechanisms of Fe(II)- and 2-oxoglutarate-dependent oxygenases. *J. Biol. Chem.* **2015**, *290*, 20702–20711. [CrossRef] [PubMed]
6. Jia, B.; Jia, X.; Kim, K.H.; Jeon, C.O. Integrative view of 2-oxoglutarate/Fe(II)-dependent oxygenase diversity and functions in bacteria. *Biochim. Biophys. Act. Gen. Subj.* **2017**, *1861*, 323–334. [CrossRef]
7. Hara, R.; Yamagata, K.; Miyake, R.; Kawabata, H.; Uehara, H.; Kino, K. Discovery of Lysine hydroxylases in the clavaminic acid synthase-like superfamily for efficient hydroxylysine bioproduction. *Appl. Environ. Microbiol.* **2017**, *83*, 1–14. [CrossRef]
8. Strieker, M.; Kopp, F.; Mahlert, C.; Essen, L.-O.; Marahiel, M.A. Mechanistic and structural basis of stereospecific Cβ-hydroxylation in calcium-dependent antibiotic, a daptomycin-type lipopeptide. *ACS Chem. Biol.* **2007**, *2*, 187–196. [CrossRef]
9. Kodera, T.; Smirnov, S.V.; Samsonova, N.N.; Kozlov, Y.I.; Koyama, R.; Hibi, M.; Ogawa, J.; Yokozeki, K.; Shimizu, S. A novel L-isoleucine hydroxylating enzyme, L-isoleucine dioxygenase from *Bacillus thuringiensis*, produces (2*S*,3*R*,4*S*)-4-hydroxyisoleucine. *Biochem. Biophys. Res. Commun.* **2009**, *390*, 506–510. [CrossRef] [PubMed]
10. Lukat, P.; Katsuyama, Y.; Wenzel, S.; Binz, T.; König, C.; Blankenfeldt, W.; Brönstrup, M.; Müller, R. Biosynthesis of methyl-proline containing griselimycins, natural products with anti-tuberculosis activity. *Chem. Sci.* **2017**, *8*, 7521–7527. [CrossRef]
11. Shibasaki, T.; Mori, H.; Chiba, S.; Ozaki, A. Microbial proline 4-hydroxylase screening and gene cloning. *Appl. Environ. Microbiol.* **1999**, *65*, 4028–4031. [CrossRef] [PubMed]
12. Koketsu, K.; Shomura, Y.; Moriwaki, K.; Hayashi, M.; Mitsuhashi, S.; Hara, R.; Kino, K.; Higuchi, Y. Refined regio- and stereoselective hydroxylation of L-pipecolic acid by protein engineering of L-proline *cis*-4-hydroxylase based on the X-ray crystal structure. *ACS Synth. Biol.* **2015**, *4*, 383–392. [CrossRef]

13. Hibi, M.; Kasahara, T.; Kawashima, T.; Yajima, H.; Kozono, S.; Smirnov, S.V.; Kodera, T.; Sugiyama, M.; Shimizu, S.; Yokozeki, K. Multi-enzymatic synthesis of optically pure β-Hydroxy α-amino acids. *Adv. Synth. Catal.* **2015**, *354*, 741–745. [CrossRef]
14. Li, P.; Wu, G. Roles of dietary glycine, proline, and hydroxyproline in collagen synthesis and animal growth. *Amino Acids* **2018**, *50*, 29–38. [CrossRef]
15. Reddy, V.P.; Kumar, A.V.; Rao, K.R. New strategy for the synthesis of *N*-aryl pyrroles: Cu-catalyzed C–N cross-coupling reaction of *trans*-4-hydroxy-L-proline with aryl halides. *Tetrahedron Lett.* **2011**, *52*, 777–780. [CrossRef]
16. Poisson, J.-F.; Orellana, A.; Greene, A.E. Stereocontrolled synthesis of (−)-kainic acid from *trans*-4-hydroxy-L-proline. *J. Org. Chem.* **2005**, *70*, 10860–10863. [CrossRef]
17. Zafar, M.I.; Gao, F. 4-Hydroxyisoleucine: A potential new treatment for type 2 diabetes mellitus. *BioDrugs* **2016**, *30*, 255–262. [CrossRef]
18. Fowden, L. Isolation of γ-hydroxynorvaline from *Lathyrus odoratus* Seed. *Nature* **1966**, *209*, 807–808. [CrossRef]
19. Broca, C.; Manteghetti, M.; Gross, R.; Baissac, Y.; Jacob, M.; Petit, P.; Sauvaire, Y.; Ribes, G. 4-Hydroxyisoleucine: Effects of synthetic and natural analogues on insulin secretion. *Eur. J. Pharmacol.* **2000**, *390*, 339–345. [CrossRef]
20. Smirnov, S.V.; Kodera, T.; Samsonova, N.N.; Kotlyarova, V.A.; Rushkevich, N.Y.; Kivero, A.D.; Sokolov, P.M.; Hibi, M.; Ogawa, J.; Shimizu, S. Metabolic engineering of *Escherichia coli* to produce (2*S*, 3*R*, 4*S*)-4-hydroxyisoleucine. *Appl. Microbiol. Biotechnol.* **2010**, *88*, 719–726. [CrossRef]
21. Zhang, C.; Li, Y.; Ma, J.; Liu, Y.; He, J.; Li, Y.; Zhu, F.; Meng, J.; Zhan, J.; Li, Z.; et al. High production of 4-hydroxyisoleucine in *Corynebacterium glutamicum* by multistep metabolic engineering. *Metab. Eng.* **2018**, *49*, 287–298. [CrossRef] [PubMed]
22. Baixue, L.; Keqiang, F.; Jian, Z.; Junjie, J.; Linjun, W.; Keqian, Y.; Yong, T. Reconstitution of TCA cycle with DAOCS to engineer *Escherichia coli* into an efficient whole cell catalyst of penicillin G. *Proc. Natl. Acad. Sci. USA* **2015**, *112*, 9855–9859.
23. Liu, L.; Hossain, G.S.; Shin, H.; Li, J.; Du, G.; Chen, J. One-step production of α-ketoglutaric acid from glutamic acid with an engineered L-amino acid deaminase from *Proteus mirabilis*. *J. Biotechnol.* **2013**, *164*, 97–104. [CrossRef] [PubMed]
24. Hossain, G.S.; Li, J.; Shin, H.; Chen, R.R.; Du, G.; Liu, L.; Chen, J. Bioconversion of L-glutamic acid to α-ketoglutaric acid by an immobilized whole-cell biocatalyst expressing L-amino acid deaminase from *Proteus mirabilis*. *J. Biotechnol.* **2014**, *169*, 112–120. [CrossRef]
25. Niu, P.; Dong, X.; Wang, Y.; Liu, L. Enzymatic production of α-ketoglutaric acid from L-glutamic acid via L-glutamate oxidase. *J. Biotechnol.* **2014**, *179*, 56–62. [CrossRef]
26. Wu, J.; Fan, X.; Liu, J.; Luo, Q.; Xu, J.; Chen, X. Promoter engineering of cascade biocatalysis for α-ketoglutaric acid production by coexpressing L-glutamate oxidase and catalase. *Appl. Microbiol. Biotechnol.* **2018**, *102*, 4755–4764. [CrossRef]
27. Schrittwieser, J.H.; Velikogne, S.; Hall, M.; Kroutil, W. Artificial biocatalytic linear cascades for preparation of organic molecules. *Chem. Rev.* **2017**, *118*, 270–348. [CrossRef]
28. Hibi, M.; Kawashima, T.; Kodera, T.; Smirnov, S.V.; Sokolov, P.M.; Sugiyama, M.; Shimizu, S.; Yokozeki, K.; Ogawa, J. Characterization of *Bacillus thuringiensis* L-isoleucine dioxygenase for production of useful amino acids. *Appl. Environ. Microbiol.* **2011**, *77*, 6926–6930. [CrossRef]
29. Hausinger, R.P. Fe(II)/α-ketoglutarate-dependent hydroxylases and related Enzymes. *Crit. Rev. Biochem. Mol. Biol.* **2004**, *39*, 21–68. [CrossRef]
30. Guterl, J.K.; Garbe, D.; Carsten, J.; Steffler, F.; Sommer, B.; Reiße, S.; Philipp, A.; Haack, M.; Rühmann, B.; Koltermann, A. Cell-free metabolic engineering: Production of chemicals by minimized reaction cascades. *ChemSusChem* **2012**, *5*, 2165–2172. [CrossRef]
31. Qiao, Z.; Xu, M.; Shao, M.; Zhao, Y.; Long, M.; Yang, T.; Zhang, X.; Yang, S.; Nakanishi, H.; Rao, Z. Engineered disulfide bonds improve thermostability and activity of L-isoleucine hydroxylase for efficient 4-HIL production in *Bacillus subtilis* 168. *Eng. Life Sci.* **2019**, *20*, 7–16. [CrossRef]
32. Hold, C.; Billerbeck, S.; Panke, S. Forward design of a complex enzyme cascade reaction. *Nat. Commun.* **2016**, *7*, 12971. [CrossRef]
33. Beer, B.; Pick, A.; Sieber, V. In vitro metabolic engineering for the production of α-ketoglutarate. *Metab. Eng.* **2017**, *40*, 5–13. [CrossRef]

34. Rose, N.R.; Mcdonough, M.A.; King, O.N.; Kawamura, A.; Schofield, C.J. Inhibition of 2-oxoglutarate dependent oxygenases. *Chem. Soc. Rev.* **2011**, *40*, 4364–4397. [CrossRef]
35. Hara, R.; Kino, K. Characterization of novel 2-oxoglutarate dependent dioxygenases converting L-proline to *cis*-4-hydroxy-l-proline. *Biochem. Biophys. Res. Commun.* **2009**, *379*, 882–886. [CrossRef]
36. Kusakabe, H.; Midorikawa, Y.; Fujishima, T.; Kuninaka, A.; Yoshino, H. Purification and properties of a new enzyme, L-glutamate oxidase, from *Streptomyces* sp. X-119-6 grown on wheat bran. *Agric. Biol. Chem.* **1983**, *47*, 1323–1328. [CrossRef]
37. Ferreira, L.A.; Pinho, S.P.; Macedo, E.A. Solubility of L-serine, L-threonine and L-isoleucine in aqueous aliphatic alcohol solutions. *Fluid Phase Equilib.* **2008**, *270*, 1–9. [CrossRef]
38. Shi, F.; Niu, T.; Fang, H. 4-Hydroxyisoleucine production of recombinant *Corynebacterium glutamicum* ssp. lactofermentumunder optimal corn steep liquor limitation. *Appl. Microbiol. Biotechnol.* **2015**, *99*, 3851–3863. [CrossRef]
39. Jing, X.; Wang, X.; Zhang, W.; An, J.; Luo, P.; Nie, Y.; Xu, Y. Highly regioselective and stereoselective hydroxylation of free amino acids by a 2-oxoglutarate-dependent dioxygenase from *Kutzneria albida*. *ACS Omega* **2019**, *4*, 8350–8358. [CrossRef]
40. Wang, X.; Nie, Y.; Xu, Y. Improvement of the activity and stability of starch-debranching pullulanase from *Bacillus naganoensis* via tailoring of the active sites lining the catalytic pocket. *J. Agric. Food Chem.* **2018**, *66*, 13236–13242. [CrossRef]
41. Xiao, R.; Anderson, S.; Aramini, J.; Belote, R.; Buchwald, W.A.; Ciccosanti, C.; Conover, K.; Everett, J.K.; Hamilton, K.; Huang, Y.J. The high-throughput protein sample production platform of the northeast structural genomics consortium. *J. Struct. Biol.* **2010**, *172*, 21–33. [CrossRef]
42. Laemmli, U.K. Cleavage of structural proteins during the assembly of the head of bacteriophage T4. *Nature* **1970**, *227*, 680–685. [CrossRef]
43. Falcioni, F.; Blank, L.M.; Frick, O.; Karau, A.; Bühler, B.; Schmid, A. Proline availability regulates proline-4-hydroxylase synthesis and substrate uptake in proline-hydroxylating recombinant *Escherichia coli*. *Appl. Environ. Microbiol.* **2013**, *79*, 3091–3100. [CrossRef]
44. Herbert, P.; Santos, L.; Alves, A. Simultaneous quantification of primary, secondary amino acids, and biogenic amines in musts and wines using OPA/3-MPA/FMOC-CI fluorescent derivatives. *J. Food Sci.* **2010**, *66*, 1319–1325. [CrossRef]
45. Sheiner, L.B.; Beal, S.L. Evaluation of methods for estimating population pharmacokinetic parameters. I. Michaelis-menten model: Routine clinical pharmacokinetic data. *J. Pharmacokinet. Biopharm.* **1980**, *8*, 553–571. [CrossRef]
46. Mori, H.; Shibasaki, T.; Uozaki, Y.; Ochiai, K.; Ozaki, A. Detection of novel proline 3-hydroxylase activities in *Streptomyces* and *Bacillus* spp. by regio- and stereospecific hydroxylation of L-proline. *Appl. Environ. Microbiol.* **1996**, *62*, 1903–1907. [CrossRef]

International Journal of
Molecular Sciences

MDPI

Article

Genetic and Functional Diversity of Nitrilases in Agaricomycotina

Lenka Rucká [1], Martin Chmátal [2], Natalia Kulik [3], Lucie Petrásková [2], Helena Pelantová [4], Petr Novotný [2], Romana Příhodová [2], Miroslav Pátek [1,*] and Ludmila Martínková [2,*]

1 Laboratory of Molecular Genetics of Bacteria, Institute of Microbiology of the Czech Academy of Sciences, Vídeňská 1083, CZ-142 20 Prague, Czech Republic; rucka@biomed.cas.cz
2 Laboratory of Biotransformation, Institute of Microbiology of the Czech Academy of Sciences, Vídeňská 1083, CZ-142 20 Prague, Czech Republic; chmatal@biomed.cas.cz (M.C.); petraskova@biomed.cas.cz (L.P.); petr.novotny@biomed.cas.cz (P.N.); romana.prihodova@biomed.cas.cz (R.P.)
3 Center for Nanobiology and Structural Biology, Institute of Microbiology of the Czech Academy of Sciences, Zámek 136, CZ-373 33 Nové Hrady, Czech Republic; kulik@nh.cas.cz
4 Laboratory of Molecular Structure Characterization, Institute of Microbiology of the Czech Academy of Sciences, Vídeňská 1083, CZ-142 20 Prague, Czech Republic; pelantova@biomed.cas.cz
* Correspondence: patek@biomed.cas.cz (M.P.); martinko@biomed.cas.cz (L.M.); Tel.: +420-296-442-398 (M.P.); +420-296-442-569 (L.M.)

Received: 24 October 2019; Accepted: 22 November 2019; Published: 28 November 2019

Abstract: Nitrilases participate in the nitrile metabolism in microbes and plants. They are widely used to produce carboxylic acids from nitriles. Nitrilases were described in bacteria, Ascomycota and plants. However, they remain unexplored in Basidiomycota. Yet more than 200 putative nitrilases are found in this division via GenBank. The majority of them occur in the subdivision Agaricomycotina. In this work, we analyzed their sequences and classified them into phylogenetic clades. Members of clade 1 (61 proteins) and 2 (25 proteins) are similar to plant nitrilases and nitrilases from Ascomycota, respectively, with sequence identities of around 50%. The searches also identified five putative cyanide hydratases (CynHs). Representatives of clade 1 and 2 (NitTv1 from *Trametes versicolor* and NitAg from *Armillaria gallica*, respectively) and a putative CynH (NitSh from *Stereum hirsutum*) were overproduced in *Escherichia coli*. The substrates of NitTv1 were fumaronitrile, 3-phenylpropionitrile, β-cyano-L-alanine and 4-cyanopyridine, and those of NitSh were hydrogen cyanide (HCN), 2-cyanopyridine, fumaronitrile and benzonitrile. NitAg only exhibited activities for HCN and fumaronitrile. The substrate specificities of these nitrilases were largely in accordance with substrate docking in their homology models. The phylogenetic distribution of each type of nitrilase was determined for the first time.

Keywords: Basidiomycota; Agaricomycotina; nitrilase; cyanide hydratase; nitrile; substrate specificity; overproduction; homology modeling; substrate docking; phylogenetic distribution

1. Introduction

Nitrilases (NLases; EC 3.5.5.-) constitute branch 1 of the nitrilase (NLase) superfamily, which comprises 13 branches of hydrolases acting on non-peptide C-N bonds [1]. NLases have attracted the attention of academia and industry due to their ability to hydrolyze nitriles under mild conditions and with stereo- or regioselectivities [2–4]. They are also promising for bioremediation, especially as far as their cyanide-degrading subtypes cyanide hydratases (CynHs; EC 4.2.1.66) and cyanide dihydratases (CynDs; EC 3.5.5.1) are concerned [5]. Our understanding of the natural roles of NLases and the cyanide-degrading enzymes still falls behind our knowledge of their biotechnological applications. Nevertheless, some probable functions of these enzymes were proposed such as nitrogen recycling, protection against herbivores, participation in the metabolism of cyano glycosides and glucosinolates, or the detoxification of hydrogen cyanide (HCN) [6].

The first NLases characterized were almost exclusively of bacterial origin [7], with some exceptions of fungal NLases, which, however, were not sequenced [8–10]. The bacterial NLases were usually classified into a number of substrate-specificity subtypes, i.e., aromatic NLases, aliphatic NLases, arylacetoNLases, bromoxynil-specific NLases, CynHs and CynDs [7]. However, a recent study suggested that aliphatic NLases may not actually exist, since the activities for aliphatic nitriles are present in other NLase types such as arylacetoNLases [11]. The plant NLases could be classified into two main subtypes, namely NIT1 through NIT3 with broad substrate specificities on the one hand [12] and NIT4 with a strict specificity for β-cyanoalanine (β-CA) on the other [13].

Recently, more than 10 new NLases in fungi of the division Ascomycota were characterized [14–16]. These enzymes could be roughly classified into aromatic NLases with a preference for (hetero) aromatic nitriles and arylacetoNLases with a preference for phenylacetonitrile (PAN) or mandelonitrile (MN). In contrast, NLases in the second largest division of fungi, Basidiomycota, have been underexplored. A single NLase was characterized in this taxon (in *Auricularia subglabra*, formerly *Auricularia delicata*) [16]. Yet understanding the function of the nitrile-converting enzymes in this taxon seems to be important with regard to the interactions of these fungi with plants. For instance, pathogenic fungi are among the factors causing damage to agricultural crops. Understanding the defense mechanisms of these fungi can lead to the identification of new targets for their suppression. In contrast, biocontrol fungi as components of the plant microbiome are known to exhibit beneficial effects on plants [17].

Basidiomycota comprise saprophytic and parasitic fungi equipped with a wealth of enzymes enabling them to degrade and utilize the plant biomass. Thus the occurrence of peroxidases, laccases, cellulases and hemicellulases in these fungi has been well documented (e.g., [18,19]). Nitrile-degrading enzymes, if active in these fungi, may support their growth on plant biomass by detoxifying natural nitriles, which are widespread in the plant kingdom [20]. For instance, toxic β-CA is formed by β-CA synthase from cysteine and HCN, whereas NIT4 transforms β-CA into a mixture of asparagine (Asn), aspartic acid (Asp) and ammonia. Thus this pathway not only serves for the detoxification of β-CA but also for nitrogen recycling. HCN, which is produced in plants during the synthesis of ethylene (phytohormone) from methionine or by cyano glycoside breakdown, can be also removed directly through its hydration catalyzed by CynHs in fungi [5].

Studies describing the transformation of nitriles or HCN in Basidiomycota are rare. Thus a psychrophilic "basidiomycete" synthesized 2-aminopropionitrile or 4-amino-4-cyanobutyric acid from HCN, ammonia and the corresponding aldehyde. The hydrolysis of these nitriles proceeded in the same organism, producing alanine and glutamic acid, respectively [21,22]. However, the enzymes catalyzing the synthesis or hydrolysis of these nitriles were not identified and the organism was not classified taxonomically. Another study reported on the elimination of CN^- in some species of Basidiomycota (*Trametes versicolor*, *Phanerochaete chrysosporium*, *Pleurotus sajor-caju*) but the mechanism of this process was not explained and no degradation products were detected except for a small amount of ammonia [23].

This work focused on the NLases in Agaricomycotina (mushroom-forming fungi). The aim of this work is to analyze their sequences, which are retrieved from the GenBank, and to predict their enzyme activities from sequence similarities to characterized NLases and from homology models. The sequences are classified into clades and representatives of three clades overproduced in *Escherichia coli* for the first time. The phylogenetic distribution of the homologues of the experimentally confirmed NLases is determined.

2. Results

2.1. Occurrence of Putative Nitrilases (NLases) in Agaricomycotina

The GenBank database was searched for putative NLases in Basidiomycota (subdivisions Agaricomycotina, Ustilaginomycotina and Pucciniomycotina). The templates were the bacterial

NLase from *Rhodococcus rhodochrous* (GenBank: ABO46008.1) [24], plant NLases NIT1 (GenBank: NP_851011.1) [12] and NIT4 (GenBank: AAM65906) [13] from *Arabidopsis thaliana*, fungal NLases from *Fusarium verticillioides* (formerly *Gibberella moniliformis*; GenBank: ABF83489.1) [15] and *Auricularia subglabra* (formerly *Auricularia delicata*; EJD42068.1) [16] and a CynH from *Aspergillus niger* (GenBank: ABX75546) [25]. The sequences which exhibited at least 25% identity (>60% cover) to at least one of the templates were retrieved. These sequences were approximately 200 in total, and the majority of them were in Agaricomycotina (around 150 sequences). If sequences exhibiting identities of ≥ 99% to each other occurred in the same species, one of them was discarded, as well as the sequences with a length atypical of NLase (≤300 or ≥400 amino acids).

Deleting the highly similar sequences and the sequences with atypical lengths resulted in obtaining a set of 135 sequences in Agaricomycotina (Supplementary Materials, Table S1). This set was used to construct a phylogenetic tree (Figure 1A) which enabled the identification of a few clades. The largest of them (clade 1; Figure 1B) consisted of 61 sequences similar to plant NLases (NIT1, NIT4) with identities of around 50%; nitrilases outside clade 1 showed lower identities to plant NLases. A smaller clade (clade 2; Figure 1C) was identified, consisting of 25 sequences with maximum ca. 50% identity to characterized NLases in Ascomycota as their closest homologues. The third clade (Figure 1D) only contained five sequences with identities of over 60% to the characterized CynH from *A. niger*. Therefore these proteins could be predicted as CynHs with a high probability and the clade was designated accordingly. Proteins outside the three clades could form an outgroup.

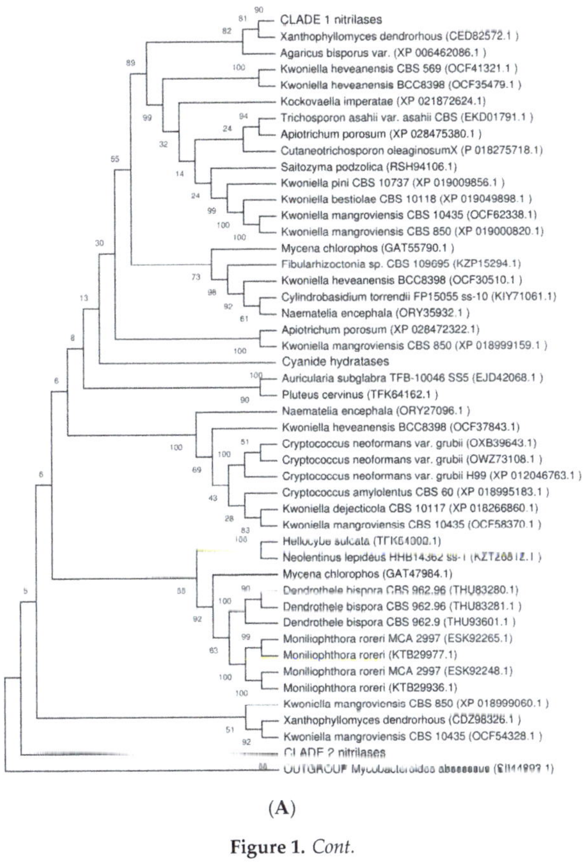

(A)

Figure 1. *Cont.*

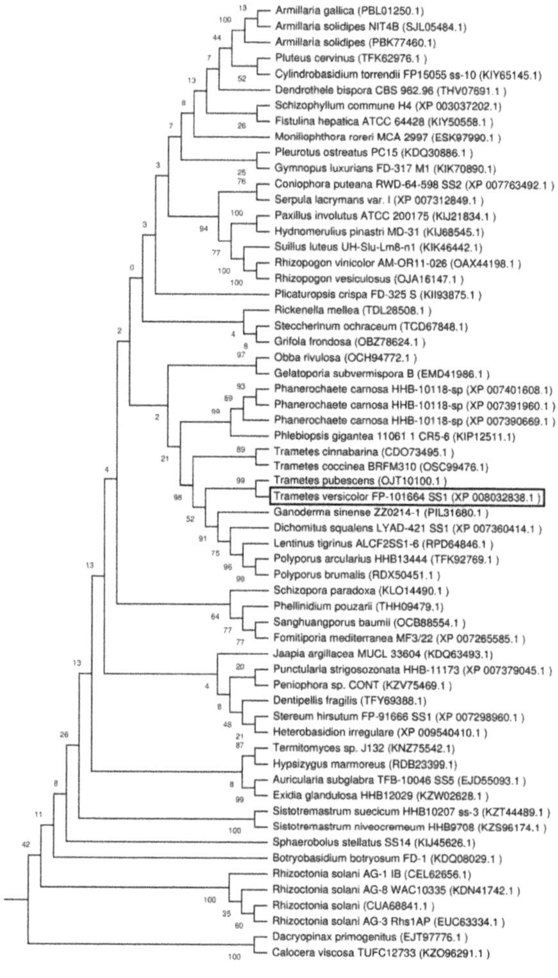

(B)

Figure 1. *Cont.*

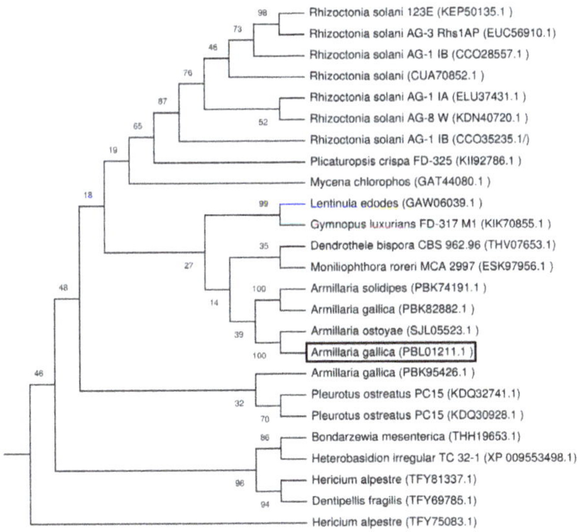

(**C**)

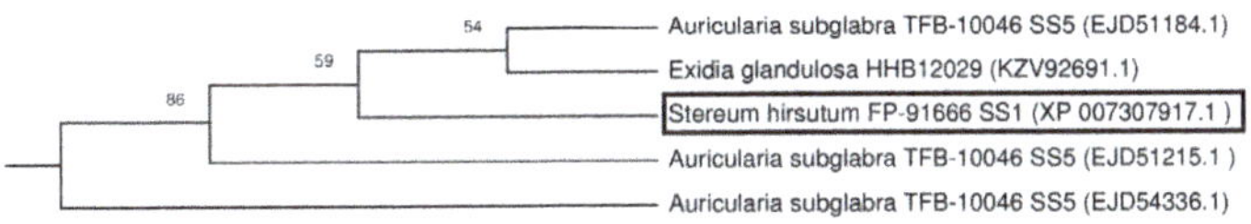

(**D**)

Figure 1. Phylogeny of nitrilases. (**A**) Full tree, (**B**) clade 1, (**C**) clade 2, (**D**) cyanide hydratases. The most parsimonious trees are shown. The consistency index is (0.340025), the retention index was (0.706964), and the composite index was 0.242788 (0.240385) for all sites and parsimony-informative sites (in parentheses). The percentage of replicate trees, in which the associated taxa clustered together in the bootstrap test (500 replicates) is shown above the branches [26]. Zero values above branches correspond to bifurcating branches and appeared as a result of tree rooting. The sequence from bacteria *Mycobacteroides abscessus* (SII44893.1) was used as the outgroup. The full tree was obtained using the subtree-pruning-regrafting (SPR) algorithm [27] with search level 1 in which the initial trees were obtained by the random addition of sequences (10 replicates). This analysis involved 135 amino acid sequences. The nitrilases (NLases) selected for overproduction are in frame.

2.2. Selection of Nitrilases for Overproduction, Multiple Alignment and Substrate Specificity Predictions

Three of the hypothetical proteins were selected for heterologous production in *E. coli*. Two of them represented the major clades of hypothetical NLases found in Basidiomycota. In addition, one of the putative CynHs was selected. These proteins originated from *Trametes versicolor* (protein NitTv1; GenBank: XP_008032838.1; clade 1), *Armillaria gallica* (protein NitAg; PBL01211.1; clade 2) and *Stereum hirsutum* (NitSh; GenBank: XP_007307917.1; CynH). The identity levels between the three proteins were below 30%.

BLAST searches in the pdb database were carried out in order to identify suitable templates that enabled us to construct three dimensional models of the selected NLases. The search revealed that all these NLases had the highest identities (30–32%) to the NLase from *Synechocystis* sp. (3wuy) [28]. In addition, NitTv1 had a 26% identity to an uncharacterized NLase-related protein 1j31 from branch 13 of the NLase superfamily (a putative apolipoprotein N-acyltransferase from *Pyrococcus horikoshii*) and to the bacterial thermophilic NLase 3ivz [29]. The 3ivz protein exhibits a high-sequence identity (82%)

to 1j31. Furthermore, the structural identity of the proteins is also high, as 75.15% of the heavy atoms of both structures can be superimposed with total distance below 1Å as calculated by Yasara GDT analysis. Therefore, 3ivz does not add any new information to fungal nitrilase sequence description and it was not used in the final sequence alignment. Lower identities were found between yeast ω-amidase designated Nit2 (4hg3) [30] and NitAg (23%) or between the pyrimidine-degrading enzyme 2vhh (EC: 3.5.1.6, branch 5) and NitSh (22%). Protein 2vhh is missing many parts of the sequence (not shown); therefore, it was discarded from further analysis. A multiple sequence alignment of the fungal NLase sequences with pdb structures was constructed (Figure 2).

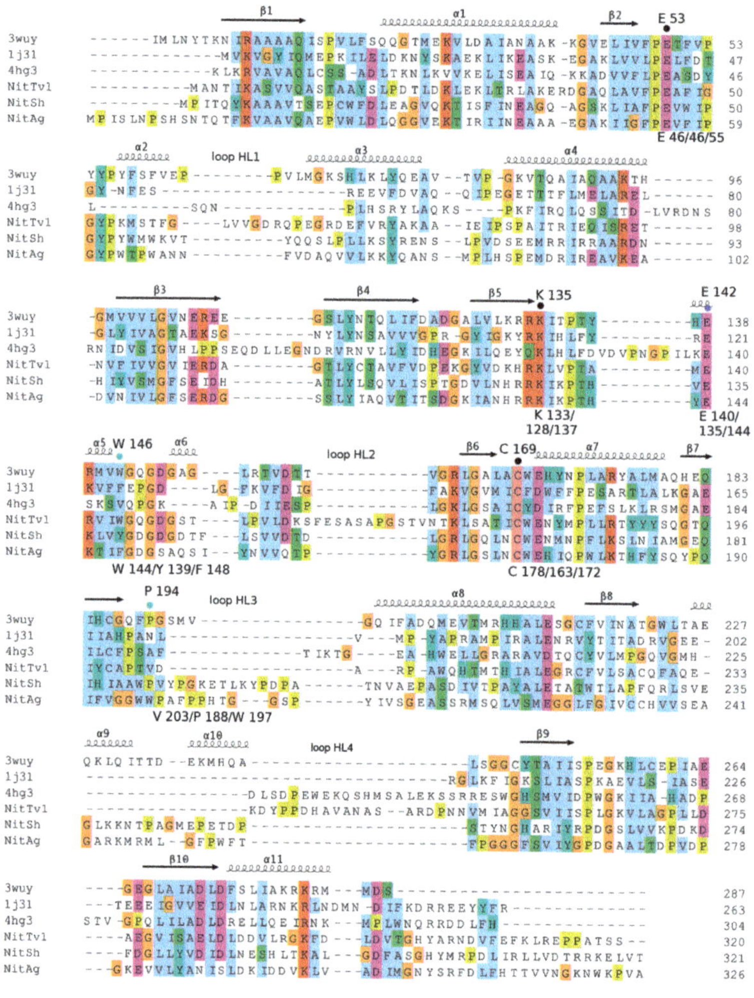

Figure 2. Multiple sequence alignment used for homology modeling. Parts of C-termini in NitAg and NitSh (compare Figure S1) were cut due to a missing template. Secondary structure elements are assigned based on 3wuy. The catalytic triad (E, K, C) is labeled with black dots, an additional E residue important for the catalytic mechanism [7] is marked with a blue dot. Other aa residues which are close to the catalytic center and could influence substrate recognition are marked with cyan dots. The numbers of aa residues above the alignment correspond to 3wuy, and those below the alignment to NitTv1/NitSh/NitAg NLases. Loops which are different in 3wuy and fungal NLases are labeled (HL1-HL4).

Previously, aligning NLases with proteins from other branches of the NLase superfamily demonstrated typical insertions in NLases [31]. Generally, the fungal NLases studied here have longer insertions (loops) than the NLase-related protein 1j31 but also NLase 3wuy. NitTv1 has longer loops, namely HL1, HL2 and HL4, while NitSh and NitAg have a longer HL3 (Figure 2). Loop HL4 in 3wuy was designated "substrate binding" and it was proposed that the larger size of this loop was the reason for the broader substrate range of the NLase from *Synechocystis* sp. in comparison with the thermophilic NLase from *Pyrococcus abyssi* [28]. This loop is similar in 3wuy and NitAg but longer in NitTv1 and NitSh.

The model of NitTv1 was built using three templates: 3wuy, 1j31 (HL3 loop) and 4hg3 (HL4 loop). The other two models were based on 3wuy alone. A few typical substrates of NLases and CynHs (**1a–8a** in Figure 3 and HCN) were docked in the models.

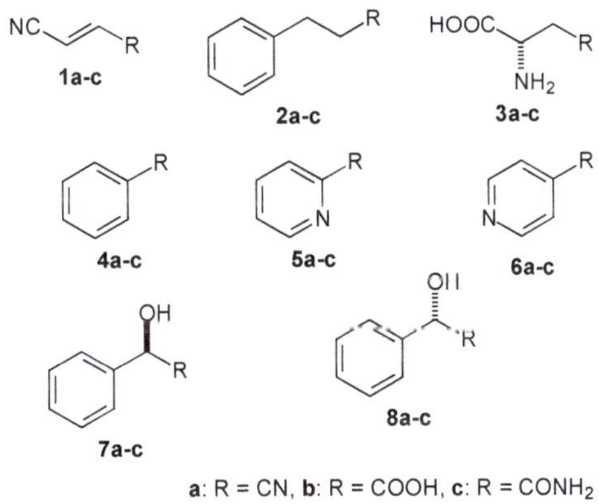

Figure 3. Nitriles examined as potential substrates of nitrilases NitTv1, NitAg and NitSh: fumaronitrile (**1a**), 3-phenylpropionitrile (**2a**), β-cyano-L-alanine (**3a**), benzonitrile (**4a**), 2-cyanopyridine (**5a**), 4-cyanopyridine (**6a**), (*R*)-mandelonitrile (**7a**), (*S*)-mandelonitrile (**8a**) and potential products of their biotransformations (carboxylic acids **1b–8b**, amides **1c–8c**).

The substrate binding in the catalytic center of the enzyme was characterized by the binding (Glide) score. Lower values of this score correspond to better binding. However, this parameter may not be sufficient to characterize the binding of small substrates such as nitriles, in which the binding score is higher due to a lack of hydrogen-bonding (HB) or stacking interactions. Thus the binding scores may be higher with nitriles than with other (larger) substrates but must not be positive. Additional geometrical constraints were also applied. Their selection was based on the analysis of the yeast ω-amidase Nit2 (4hg3)-oxaloacetate (reaction product) complex (4hg5) [30], and on the NLase catalytic mechanism [32] in which the catalytic E forms a HB with the cyano nitrogen. This N-atom must be positioned within HB distance of the catalytic K, and also close to the catalytic E [32,33]. Furthermore, the correct orientation of the substrate was characterized by the distance of the C-atom in the CN group from the S-atom in the catalytic C (optimally < 3.1 Å). The broadest substrate specificity was predicted for NitSh. In its model, the Glide scores suggested its possible activities for fumaronitrile (FN) (**1a**), β-CA (**3a**), benzonitrile (BN) (**4a**), 2-cyanopyridine (2CP) (**5a**), (*R*)-MN (**7a**), (*S*)-MN (**8a**) and HCN. The distance between the C-atom of the substrate and the S-atom in the catalytic C indicated a suitable orientation of these substrates, being below or only slightly above 3.1 Å. The model of NitTv1 exhibited suitable Glide scores and substrate orientation with FN, 3-phenylpropionitrile (PPN) (**2a**) and β-CA. In NitAg, these docking parameters were only suitable with FN and HCN (Table 1).

Table 1. Docking of selected substrates in fungal NLase models.

Proposed Substrate	Docking Glide Score [kcal/mol]		
	NitTv1	NitAg	NitSh
Fumaronitrile (**1a**)	−0.15	−1.0	−0.77
3-Phenylpropionitrile (**2a**)	−3.49	– [1]	n.i.
β-Cyano-L-alanine (**3a**)	−2.82	n.i.	−2.60
Benzonitrile (**4a**)	n.i.	n.i.	−4.56
2-Cyanopyridine (**5a**)	n.i.	n.i.	−4.22
4-Cyanopyridine (**6a**)	n.i.	n.i.	n.i.
(*R*)-Mandelonitrile (**7a**)	n.i.	n.i.	−5.18
(*S*)-Mandelonitrile (**8a**)	n.i.	n.i.	−4.78
Hydrogen cyanide (HCN)	n.i.	−3.50	−3.12

[1] In this case, the substrate was docked in the active site close to C, but the nitrogen in the cyano group of the substrate is rotated outwards from the catalytic E and K, different from the proposed orientation of the intermediate in the 4hg5 crystal [30]. n.i. = not identified. In this case, the program Glide was unable to dock the substrate with the defined geometrical constraints (see text).

2.3. Modeling Substrate Interaction with the Active Site Residues

The catalytic residues E, K, C interact by HB and the second E residue participates in Lys stabilization by a salt bridge (Figure 4 and Supplementary materials, Table S2). Another residue found in a close contact with the ligand corresponds to W146 in 3wuy (W144, F148 and Y139 in NitTv1, NitAg and NitSh, respectively; Figure 4, Figure 5 and Figure S2).

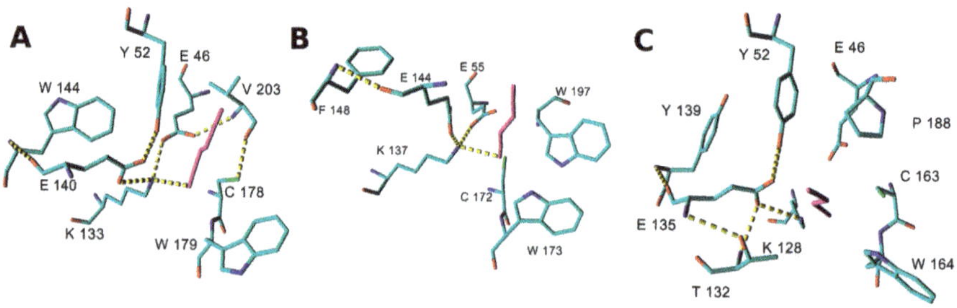

Figure 4. Snapshots of the complexes of nitrilase (**A**) NitTv1, (**B**) NitAg and (**C**) NitSh with fumaronitrile after 5 ns of molecular dynamics simulations. Hydrogen bonds (HBs) are shown with yellow dashed lines, fumaronitrile is colored magenta, protein residues have element color, and hydrogens are hidden. Amino acids within 3 Å from ligand and those marked with dots in Figure 2 are shown; some residues are hidden for clarity.

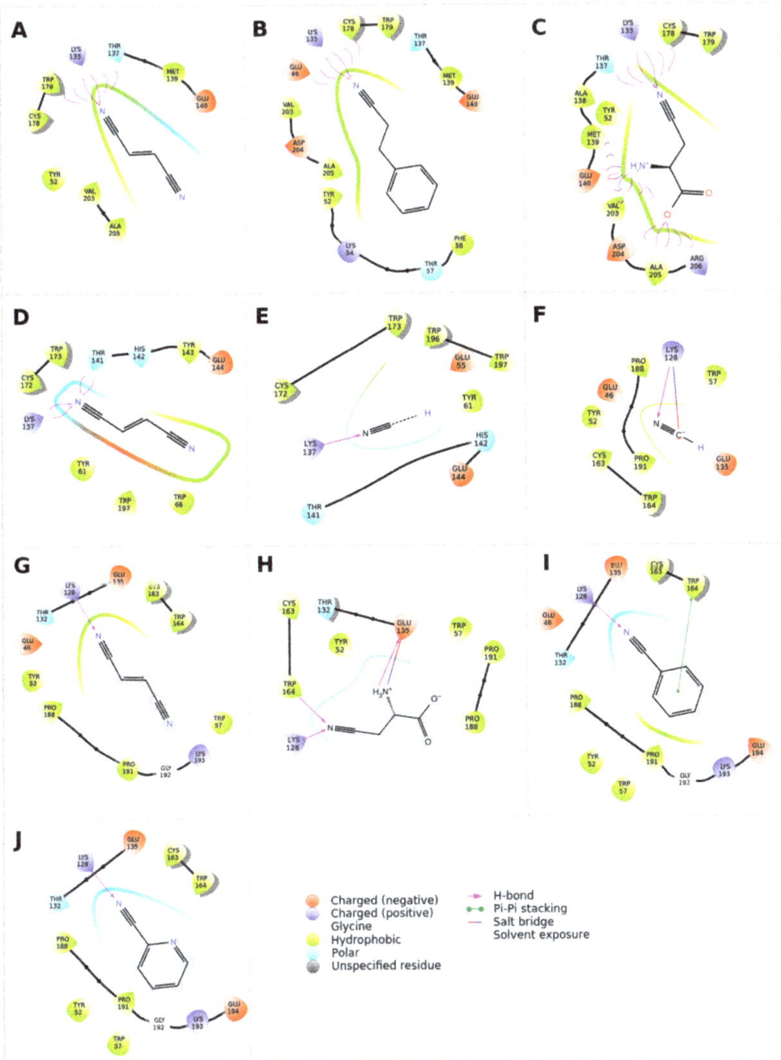

Figure 5. Orientation of ligands docked in the active site of nitrilases NitTv1 (**A–C**), NitAg (**D**,**E**) and NitSh (**F–J**). Enzyme residues within 3.9 Å from ligands are shown as raindrops and colored according to their properties (see legend in the right bottom angle). Ligands are represented as stick model; carbons and hydrogens are hidden: (**A**,**D**,**G**) fumaronitrile; (**B**) 3-phenylpropionitrile; (**C**,**H**) β-cyano-L-alanine, (**E**,**F**) HCN; (**I**) benzonitrile; (**J**) 2-cyanopyridine.

The corresponding Y142 in the NLase from *Rhodococcus rhodochrous* was proposed to be important for the substrate specificity of the enzyme: the replacement of an aromatic residue with a non-aromatic one (A, V or L) abolished the enzyme activity for aliphatic nitriles [33]. All fungal NLases examined here have an aromatic residue at this position except for the yeast Nit 2 protein (pdb code 4hg3). This residue is not directly in the active site, but close to the catalytic E (E140, E144 and E135 in NitTv1, NitAg and NitSh, respectively; Figure 4) and it may stabilize its correct orientation through a HB formed by its backbone atoms. The exact mechanism of discrimination of substrate affinity by aromaticity of this residue is not clear. However, mutation to a smaller A residue may allow more space for catalytic E side chain rotation.

Another residue which could influence the substrate recognition corresponds to P194 in 3wuy (V203, W197 and P188 in NitTv1, NitAg and NitSh, respectively; Figures 4 and 5). P194 is not conserved in NLases. In 3wuy it was identified as one of the hydrophobic residues in the "substrate-binding loop" [28]. W197 together with F200 in NitAg narrows the binding site, suggesting that the enzyme will not accept big substrates such as PPN, in contrast to NitTv1 (Supplementary materials, Figure S2). Most of the sequences in clade 2 (similar to NitAg) have a W residue in the corresponding position. From other clades only KII93875.1 from *Plicaturopsis crispa* (assigned to clade 1) has W in this position, while other clade 1 sequences have mostly a nonpolar (hydrophobic) V, A or P substitution. Clade 2 sequences share conserved WWP (residues 196–198 in NitAg) motif and loop HL3. NitSh has WPV residues in the corresponding position. Hence we could conclude that a mutation at this point could modulate the substrate affinity: smaller residues should allow bigger substrates.

The residue one position downstream of the catalytic C (W179, W173 and W164 in NitTv1, NitAg and NitSh, respectively) is also close to the substrate (Figures 4 and 5). The majority of nitrilases in Ascomycota also have a W residue at this position [34] like 3wuy. The aromatic residue at this position seems to be important for the charge stabilization upon catalysis and participates in T-shaped stacking with aromatic substrates (Figure 5I). However there are some sequences in the dataset (see Supplementary materials, Table S1), which have a CGE instead of the CWE motif, e.g., in KIY71061.1 (*Cylindrobasidium torrendii*), KZP 15294.1 (*Fibularhizoctonia* sp.), OCF30510.1 (*Kwoniella heveanensis*), ORY35932.1 (*Naematelia encephala*), or other residues after catalytic C, e.g., in ELU37431.1 (*Rhizoctonia solani*), XP_012046763.1, OXB39643.1 and OWZ73108.1 (*Cryptococcus neoformans*), XP_007401608.1 (*Phanerochaete carnosa*) and THH19653.1 (*Bondarzewia mesenterica*). The sequence from *P. carnosa* belonged to clade 1 and those from *R. solani* and *B. mesenterica* to clade 2 but the other ones were not classified into any of these clades.

2.4. Determination of Nitrilase Activities in E. coli Cells

Isopropyl β-D-1-thiogalactopyranoside (IPTG) induced whole cells were examined for their activities towards typical NLase and CynH substrates which were also used for docking (see above). Preliminary screening of the substrates was carried out using high-performance liquid chromatography (HPLC) or spectrophotometric methods suitable for determining the expected products (see Materials and Methods). Thus the reaction mixtures from the transformations of substrates **1a**–**6a** (Figure 3) and HCN were checked for the presence of the corresponding acids or amides by HPLC. The retention times and ultraviolet/visible (UV/vis) spectra of the products were compared with those of authentic compounds **1b**–**6b**, **1c**, **3c**, **5c** and **6c** (commercial standards except for **1b** and **1c**, which were prepared in our previous work) [16]. The reaction mixtures from the transformations of HCN were examined for the production of formamide, which was determined spectrophotometrically and by HPLC. The production of ammonia from β-CA was determined spectrophotometrically.

The reaction products were detected for FN, PPN, 4CP and β-CA in NitTv1, for FN, BN, 2CP and HCN in NitSh, and for FN and HCN in NitAg. No products or only traces of products were determined in other cases. The identity of the products was then confirmed by NMR (acids **1b**, **3b**; amides **1c**, **2c**) and/or LC-MS (acids **1b**, **2b**, **5b**, **6b**; amides **3c**, **5c**, **6c**) (Supplementary materials, Figures S3–S13). Benzoic acid (**4b**) was identified by comparing its retention time and UV spectrum to those of the authentic standard (λmax 228.7 nm).

The major products were largely carboxylic acids but amides were produced in significant amounts in some cases. Thus amide (formamide) was produced as the major product from HCN by NitSh and NitAg. NitTv1 transformed β-CA (**3a**) into a mixture of Asp (**3b**) and Asn (**3c**) at a ratio of 72:28. Significant amounts of amides **5c** and **6c** were also found in the products from 2CP (**5a**) and 4CP (**6a**) obtained with NitSh and NitTv1, respectively. The transformation of FN (**1a**) largely proceeded on only one of the cyano groups. The ratio of acid **1b** and amide **1c** depended on the origin of the enzyme. NitTv1 produced more cyano amide **1c** than cyano acid **1b** in contrast to the other two enzymes. The products of NitAg were only detected with high cell densities (OD_{600} of ca. 10). NitTv1 also formed a minor product **1d** (Figure S3) and traces of fumaric acid (not shown) as a result of the transformation of both cyano groups in FN.

The specific activities for the aforementioned substrates were quantified from the rates of product formation determined by HPLC or spectrophotometry (Table 2). FN was the preferred substrate of NitTv1, while β-CA, PPN and 4CP were transformed at much lower rates (< 8%). The most favored substrate of NitSh was HCN, followed by 2CP, FN and BN transformed at rates one to two orders of magnitude lower. NitAg also exhibited its highest activity for HCN but this activity was almost 50 times lower than at NitSh.

Table 2. Relative activities of the selected nitrilases.

Substrate	Relative Activity [%]		
	NitTv1	NitAg	NitSh
Fumaronitrile (**1a**)	100 [1]	0.140 ± 0.003	0.199 ± 0.003
3-Phenylpropionitrile (**2a**)	6.26 ± 0.13	0	0
β-Cyano-L-alanine (**3a**)	7.96 ± 1.57	0	traces
Benzonitrile (**4a**)	0	0	0.135 ± 0.010
2-Cyanopyridine (**5a**)	traces	0	1.21 ± 0.03
4-Cyanopyridine (**6a**)	5.26 ± 0.26	traces	traces
HCN	0	100 [2]	100 [3]

[1] Activity of 2.30 ± 0.06 U/mg dry cell weight was taken as 100%. [2] Activity of 3.22 ± 0.50 U/mg dry cell weight was taken as 100%. [3] Activity of 153 ± 20 U/mg dry cell weight was taken as 100%.

The experimentally determined activities were in accordance with the predictions, however with some exceptions. NitSh exhibited the broadest substrate specificity (for HCN, 2CP, FN and BN) in accordance with the docking experiments. Nevertheless, the activity predicted for β-CA was not confirmed in this enzyme. The preference of NitTv1 for β-CA, FN, PPN and 4CP and NitAg for HCN corroborated the predictions. However, a significant discrepancy between the docking experiments and the activity assays was observed with MN (the former indicated its suitable orientation in the active sites of NitSh and NitTv1 but no activity was determined in the latter). With NitSh, this may be explained by the racemization of MN resulting in the formation of HCN and benzaldehyde, followed by the removal of HCN by this enzyme. Therefore, MN cannot be restored and benzaldehyde remains in the reaction mixture.

2.5. Phylogenetic Distribution of NLases in Basidiomycota

In the subdivision Agaricomycotina, NLase sequences were found in 18 orders and 64 genera (Table 3). The order with the highest number of NLases (38) was Agaricales, followed by Tremellales and Polyporales with 21 and 16 NLase sequences, respectively. The highest numbers of NLase sequences were found in the genera *Kwoniella* (13), *Rhizoctonia* (11), *Armillaria* (8) and *Moniliophthora* (6).

Sequences of clade 1 were found in 13 orders, while those of clade 2 only occurred in four orders. Four orders contained sequences of both clades. Sequences of clade 1 were most frequent in the orders Agaricales and Polyporales with 14 and 16 sequences, respectively. These sequences occurred in the classes Agaricomycetes and Dacrymycetes but not in the class Tremelomycetes (orders Tremellales and Cystofilobasidiales). Clade 2 NLases were only present in the classes Agaricomycetes and they most often occurred in the orders Agaricales (13) followed by Canthareles (seven; all of them in genus *Rhizoctonia*) and Russulases (five). CynHs were also only found in *Agaricomycetes*, and they were quite rare, only occurring in three genera, i.e., *Auricularia*, *Exidia* (Auriculariales) and *Stereum* (Russulales). Sequences not classified into the aforementioned clades were mainly found in Tremellales (21) and Agaricales (10).

Table 3. Phylogenetic distribution of nitrilases in Basidiomycota.

Order [1] [Number of Sequences]	Genus [1] [Number of Sequences]		
	Clade 1	**Clade 2**	**Others**
Agaricales (38)	*Armillaria* (3) *Cylindrobasidium* (1) *Dendrothele* (1) *Fistulina* (1) *Gymnopus* (1) *Hypsizygus* (1) *Lentinus* (1) *Moniliophthora* (1) *Pleurotus* (1) *Pluteus* (1) *Schizophyllum* (1) *Termitomyces* (1)	*Armillaria* (5) *Dendrothele* (1) *Gymnopus* (1) *Lentinus* (1) *Moniliophthora* (3) *Mycena* (1) *Pleurotus* (2)	*Agaricus* (1) *Cylindrobasidium* (1) *Dendrothele* (3) *Pluteus* (1) *Moniliophthora* (2) *Mycena* (2)
Amylocorticiales (2)	*Plicaturopsis* (1)	*Plicaturopsis* (1)	
Atheliales (1)			*Fibulanrhizoctonia* (1)
Auriculariales (7)	*Auricularia* (1) *Exidia* (1)		*Auricularia* (4 [2]) *Exidia* (1 [3])
Cystofilobasidiales (2)			*Xanthophylomyces* (2)
Boletales (7)	*Coniophora* (1) *Hydnomerulius* (1) *Paxillus* (1) *Rhizopogon* (2) *Serpula* (1) *Suillus* (1)		
Cantharellales (12)	*Botryobasidium* (1) *Rhizoctonia* (4)	*Rhizoctonia* (7)	
Corticiales (1)	*Punctularia* (1)		
Dacrymycetales (2)	*Calocera* (1) *Dacryopinax* (1)		
Geastrales (1)	*Sphaerobolus* (1)		
Gloeophyllales (2)			*Neolentinus* (1) *Heliocybe* (1)
Hymenochaetales (5)	*Fomitiporia* (1) *Phellinidium* (1) *Rickenella* (1) *Sanghuangporus* (1) *Schizopora* (1)		
Jaapiales (1)	*Jaapia* (1)		
Polyporales (16)	*Dichomitus* (1) *Ganoderma* (1) *Gelatoporia* (1) *Grifola* (1) *Obba* (1) *Phanerochaete* (3) *Phlebiopsis* (1) *Polyporus* (2) *Steccherinum* (1) *Trametes* (4)		
Russulales (10)	*Dentipellis* (1) *Heterobasidion* (1) *Peniophora* (1) *Stereum* (1)	*Bondarzewia* (1) *Dentipellis* (1) *Hericium* (2) *Heterobasidion* (1)	*Stereum* (1 [3])
Trechisporales (2)	*Sistotremastrum* (2)		
Tremellales (21)			*Cryptococcus* (4) *Kockovaella* (1) *Kwoniella* (13) *Naematelia* (2) *Saitozyma* (1)
Trichosporonales (4)			*Apiotrichum* (2) *Cutaneotrichosporon* (1) *Trichosporon* (1)

[1] Assigned according to https://www.ncbi.nlm.nih.gov/Taxonomy. [2] three cyanide hydratases and an arylacetonitrilase. [3] cyanide hydratase. Note: Orders belonging to the class Tremellomycetes and Dacrymycetes are highlighted in light grey and dark grey, respectively. Other orders belong to class Agaricomycetes.

3. Discussion

Basidiomycota are comprised of almost 30,000 known species with 70% of them in their largest subdivision, Agaricomycotina, which, in turn, has Agaricomycetes as its largest class (the smaller ones are Dacrymycetes and Tremellomycetes, together comprising only ca. 2% of the species [35]). In accordance with the sizes of these classes, the searches for putative NLases in Basidiomycota revealed that the majority of these enzymes occurred in the class Agaricomycetes. The study focused on the analysis of a set of 135 non-redundant putative NLases from Agaricomycotina, aiming to elucidate the function and taxonomical distribution of this type of protein.

In this NLase set, a few phylogenetic clades could be distinguished, first of all the two largest clades with 61 and 25 sequences. The distribution of the clades seemed to be taxon-specific. According to the current status of the database, clade 1 occurred in Agaricomycetes and Dacrymycetes, while clade 2 was only present in Agaricomycetes. Putative CynHs were also only found in this class. Tremellomycetes contained a few putative NLases, but these proteins did not belong to any of the aforementioned clades.

To shed light on the activities and functions of the putative fungal NLases from Agaricomycotina, two approaches were used–modeling and substrate docking on one hand and heterologous gene expression followed by activity assays on the other. The results of both approaches were in good accordance with only a few exceptions.

These strategies enabled us, with some degree of certainty, to assign the substrate specificities to the largest NLase clades in Agaricomycotina. The closest characterized homologue of NitTv1 (representative of clade 1) was the NLase NIT4 from *Arabidopsis thaliana* (GenBank: AAM65906) with 52% identity (98% cover), but NIT1, NIT2 and NIT3 NLase isoforms from the same organism (GenBank: NP_001078234.1, CAA68934.3 and NP_190018.1, respectively) also shared relatively high identities (47–49%, 98–99% cover) to NitTv1. NIT1-NIT3 on one hand, and NIT4 on the other have distinct substrate specificities: NIT4 was found to have β-CA (an intermediate of cyanide assimilation [6]) as its preferential substrate and hydrolyzed PPN (product of glucosinolate metabolism in plants [6]) with activities more than 100 times lower than those for β-CA [13]. In contrast, NIT1-3 operated on different substrates, primarily arylaliphatic nitriles such as PPN, whereas the conversion of β-CA was less than 1% compared to PPN [12]. Due to these differences between NIT1-NIT3 and NIT4, it was not possible to predict the substrate specificity of NitTv1 on the basis of sequence similarities. Whole-cell assays suggested that NitTv1 was active for both β-CA and PPN (in accordance with the docking experiments) unlike both types of plant NLases. In addition, NitTv1 transformed FN, a substrate not examined with NIT4 [13] or NIT1-NIT3 [12], with activities ca. 12.5 times higher compared to β-CA.

The similarity of NLases from clade 1 to plant NLases suggests that Basidiomycota could obtain the corresponding genes from plants by horizontal gene transfer (HGT). NIT4 is ubiquitous in plants, in contrast to NIT1-3, which is typical for Brassicaceae [12]. It may be hypothesized that the clade 1 NLases in Basidiomycota evolved from the widespread NIT4 enzymes and acquired new substrate specificities and functions. The NLases of clade 1 are the most widespread within Basidiomycota. They have been most frequently found in Agaricales and Polyporales. The transfer of the genes between plants and Basidiomycota is plausible due to frequent interactions between these organisms. For instance, many members of the order Agaricales are symbionts, saprophytes or pathogens of plants, and Polyporales are abundant in wood-rotting species (saprophytes but also pathogens). *Rhizoctonia solani* (Cantharellales), a species with the highest number of NitTv1 homologues, is a plant pathogen able to infect a number of various hosts such as rice, corn, sorghum, bean, soybean, cabbage, lettuce and many others [36].

The closest characterized homologues of NitAg, which was selected as a representative of clade 2, are various NLases from Ascomycota [15,16] with maximum ca. 50% identity to NitAg. Of the tested substrates, only HCN and FN were transformed by this enzyme. The major reaction product of HCN was formamide.

The NitAg homologues are much less abundant in Basidiomycota than NitTv1 homologues and mainly occur in a few genera of the orders Agaricales and Russulales, and in *Rhizoctonia solani* (Cantharellales). In particular *R. solani* is rich in NitAg homologues, which may be due to the exposure of the pathogen to cyanides produced by the plant host. The hypothesis of HGT may be also applied to the NLases in clade 2. It is possible that Basidiomycota acquired these genes from Ascomycota, which co-infect the same host. In fact the infection of a plant host by multiple pathogenic species is common [37].

Similarly, HGT is possible with CynHs, which rarely occur in Basidiomycota but are frequent in Ascomycota, in which a number of CynHs originating from, e.g., *Gloeocercospora sorghi*, *Neurospora crassa*, *Gibberella zeae* or *Aspergillus nidulans* were expressed in heterologous hosts such as *Aspergillus nidulans* or *E. coli* [38,39]. Furthermore, sequenced genomes of Ascomycota contain numerous putative CynHs. In contrast, only five CynHs were identified in Basidiomycota so far (in Auriculariales and Russulales, i.e., saprophytes but also plant pathogens such as *S. hirsutum*). In this case, the prediction of these enzymes as CynHs was possible due to highly conserved primary sequences (with identities of over 60% to characterized CynHs). This assumption was confirmed by activity assays with NitSh from *S. hirsutum*. Its substrate specificity for FN and 2CP, in addition to HCN, was similar to those of some previously characterized CynHs from e.g., *A. niger*, *Botryotinia fuckeliana* or *Pyrenophora teres* [16,25]. NitSh was most similar to the CynH from *N. crassa* (84.5% identity, 94% cover), which was partly characterized previously and was found to be the most stable CynH of those tested [39]. Therefore, the utility of NitSh in the bioremediation of cyanide-containing wastes will be examined.

FN was a substrate of all three of the examined enzymes. This is in accordance with our previous finding that FN was hydrolyzed by all tested NLases of different types and origin [16]. However, the products were dependent on the enzyme. NitTv1 mainly produced the corresponding cyano amide like the previously examined NLase from *Aspergillus kawachii*, which was classified as an "aromatic NLase" [16]. The major product of FN transformation by NitSh was cyano acid instead. This reaction pattern was similar to that of the NLase NitMp from *Macrophomina phaseolina* [16].

In summary, the genomes of Basidiomycota contain genes encoding functional NLases, with the majority of them occurring in the subdivision Agaricomycotina, class Agaricomycetes. The NLases in Agaricomycotina were classified into clades with different substrate specificities and different taxonomical distributions. Three of the enzymes were for the first time characterized in terms of their substrate specificities. Two of them (from *T. versicolor* and *S. hirsutum*) were similar to NLases from other organisms, namely NLases from plants and CynHs from Ascomycota. The third enzyme (from *A. gallica*) had no close characterized homologues among NLases from other organisms. NLases in Basidiomycota need to be studied further with focus on the regulation of their production in fungi. This should help to better understand and exploit the bioremediation potential of these fungi.

4. Materials and Methods

4.1. Sequence Analysis

The GenBank database was searched for NLase sequences using the program BLAST (Available online: http://blast.ncbi.nlm.nih.gov/Blast.cgi) with published NLase sequences (see section Results) as templates. For phylogenetic analysis, multiple sequence alignment was built with the MUSCLE algorithm [40]. Non-informative sites were deleted, i.e., the N- and C-terminal parts were truncated and the regions with less than 90% site coverage were eliminated during alignment preparation in MEGA X [41]. There were a total of 282 positions in the final dataset. Evolutionary analyses were conducted in MEGA X and used for phylogenetic tree building with the maximum parsimony method (500 bootstraps). The cut-off parameter used to isolate the clades from the global tree was a bootstrap value of ≥ 90%. The taxonomical classification of the strains in which NLases were sequenced was according to available online: https://www.ncbi.nlm.nih.gov/Taxonomy.

4.2. Nitrilase Modeling and Substrate Docking

BLAST was used to search for homologous structures in the Protein Data Bank (pdb) database [42]. Multiple sequence alignments were carried out using the program T-Coffee [43] and manually corrected in Jalview [44] based on structure alignment with Yasara [45]. Sequences with the highest identities to the examined NLases were selected as templates for homology modeling with MODELLER 9.16 [46]. The observed models were validated with the structure validation tool MOLPROBITY [47] and minimized using a water model of TIP3P type in Yasara. Structures of substrates were downloaded from the Pubchem database (Available online: https://pubchem.ncbi.nlm.nih.gov/) and minimized with the LigPrep tool in the software Schrödinder [48]. The docking of potential substrates was done using the program Glide (Schrödinger, LLC).

4.3. Nitrilase Overproduction

The synthesis of the NLase genes, which were optimized for *E. coli*, was performed with GeneArt (ThermoFisher Scientific, Regensburg, Germany). The optimized gene sequences are shown in Supplementary Material—Gene and protein sequences (Figure S1). Gene cloning and strain cultivation were as described previously [16]. Briefly, the expression vector was constructed by cloning the gene into the *NdeI* and *XhoI* sites of pET22b(+) and used to transform *E. coli* Origami B (DE3) co-expressing the GroEL/ES chaperone. The cultures were grown in 2 × YT medium and the induction of NLase and chaperone gene expression was performed with 0.02 mM IPTG and 0.011 mM L-arabinose, respectively, at 20 °C for 24 h. The cells were centrifuged and used for enzyme assays.

4.4. Nitrilase Activity Assays

The assays were carried out as described previously [16] with minor modifications. Eppendorf tubes (1.5-mL) contained 0.5 mL of the reaction mixture consisting of 50 mM Tris/HCl buffer, pH 8.0, 150 mM NaCl, 25 mM substrate, 5% (*v*/*v*) methanol (cosolvent) and an appropriate amount of whole cells. The substrate was BN, PPN, FN or MN. The reactions with 2CP, 4CP, β-CA or free cyanide (KCN) as substrates were carried out analogously but without methanol. The reactions proceeded at 30 °C with shaking, and were terminated after 5–10 min by the addition of 0.050 mL of 2 M HCl for nitriles except β-CA. The reaction of β-CA was terminated by centrifugation of the cells. The reaction of HCN was terminated by adding 1.0 mL of methanol and centrifugation. The activities for β-CA were calculated from the production of ammonia, which was determined spectrophotometrically [11,49]. The activities for HCN were determined from the production of formamide, which was determined spectrophotometrically [50] and by HPLC (see below). The activities for other substrates were determined from the production of carboxylic acids and/or amides, which were determined by HPLC (see below).

4.5. Analytical High-Performance Liquid Chromatography (HPLC)

The concentrations of PPN, BN and MN, and their reaction products were determined using a Chromolith SpeedRod RP-18 column (50 mm × 4.6 mm; Merck KgaA, Darmstadt, Germany) and acetonitrile (ACN)/water (20/80, *v*/*v*) with 0.1% phosphoric acid as mobile phase at a flow rate of 2 mL/min. The concentrations of FN, 2CP and 4CP, and their reaction products were determined using an ACE C8 column (5 μm, 250 mm × 4 mm; Advanced Chromatography Technologies Ltd., Aberdeen, U.K.) with ACN/5 mM sodium phosphate buffer, pH 7.2 (10/90, *v*/*v*) at a flow rate of 0.9 mL/min. Formamide was determined by hydrophilic interaction HPLC [51] under modified conditions using a Nucleosil silica column (5 μm, 250 mm × 4.6 mm; Hichrom, Lutterworth, U.K.) with ACN/water (90/10, *v*/*v*) at a flow rate of 1 mL/min. The column temperature was 34 °C in all methods. The retention times and UV spectra of the substrates and products were in accordance with those of the authentic standards.

4.6. Liquid Chromatography–Mass Spectrometry (LC–MS) Analysis

The reaction mixtures were analyzed using a Shimadzu Prominence system consisting of a DGU-20A$_3$ mobile phase degasser, two LC-20AD solvent delivery units, a SIL-20AC cooling autosampler, a CTO-10AS column oven, an SPD-M20A diode array detector and a liquid chromatography–mass spectrometry (LC–MS-2020) mass detector with a single quadrupole equipped with an electrospray ion source (Shimadzu Europa GmbH, Duisburg, Germany). The column was an ACE 5 C8 (5 µm, 250 mm × 4 mm; Advanced Chromatography Technologies Ltd., Aberdeen, U.K.) or Chromolith RP 18e (100 mm × 3 mm; Merck KgaA, Darmstadt, Germany) and the mobile phase consisted of ACN/water (10/90). Analysis was performed at a flow rate of 0.4 mL/min and a column temperature of 34 °C. The MS parameters were as follows: negative mode; ESI interface voltage, 4.5 kV; detector voltage, 1.15 kV; nebulizing gas flow, 1.5 mL/min; drying gas flow, 15 mL/min; heat block temperature, 200 °C; DL temperature, 250 °C; SCAN mode 50–250 *m*/*z*; software LabSolutions version 5.75 SP2.

4.7. Nuclear Magnetic Resonance (NMR) Analysis

Nuclear magnetic resonance (NMR) analysis was performed on Bruker AVANCE III 400 and 700 MHz spectrometers (Bruker BioSpin, Rheinstetten, Germany) in deuterated dimethylsulfoxide (DMSO-d_6) or D_2O at 30 °C. Residual signal of solvents (DMSO-d_6: δ_H 2.499 ppm, δ_C 39.46 ppm; D_2O: δ_H 4.508 ppm) served as internal standards; the carbon spectra in D_2O were referenced to the acetone signal (δ_C 30.50). NMR experiments were performed using standard manufacturers' software.

Supplementary Materials: Supplementary materials can be found at http://www.mdpi.com/1422-0067/20/23/5990/s1.

Author Contributions: Conceptualization, M.P. and L.M.; Data curation, N.K. and H.P.; Funding acquisition, M.P. and L.M.; Investigation, L.R., M.C., N.K., L.P., H.P., P.N. and R.P.; Methodology, L.P., H.P., M.P. and L.M.; Project administration, M.P. and L.M.; Supervision, M.P. and L.M.; Writing—original draft, N.K. and L.M.; Writing—review and editing, L.R., N.K., M.P. and L.M.

Funding: This research was funded by the Czech Science Foundation, grant number 18-00184S.

Conflicts of Interest: The authors declare no conflict of interest.

Abbreviations

ACN	Acetonitrile
BN	Benzonitrile
β-CA	β-Cyano-L-alanine
2CP	2-Cyanopyridine
4CP	4-Cyanopyridine
CynH	Cyanide hydratase
CynD	Cyanide dihydratase
DL	Desolvation line
DMSO-d_6	Deuterated dimethylsulfoxide
ESI	Electrospray ionization
FN	Fumaronitrile
GDT	Global Distance Test
HB	Hydrogen bond
HGT	Horizontal gene transfer
IPTG	Isopropyl β-D-1-thiogalactopyranoside
MN	Mandelonitrile
MUSCLE	Multiple Sequence Comparison by Log-Expectation
NitAg	Nitrilase from *Armillaria gallica*
NitSh	Nitrilase from *Stereum hirsutum*

NitTv1	Nitrilase from *Trametes versicolor*
NLase	Nitrilase
PAN	Phenylacetonitrile
PPN	3-Phenylpropionitrile
RP	Reverse phase

References

1. Pace, H.C.; Brenner, C. The nitrilase superfamily: Classification, structure and function. *Genome Biol.* **2001**, *2*, reviews0001-1. [CrossRef] [PubMed]
2. Gong, J.-S.; Lu, Z.-M.; Li, H.; Shi, J.-S.; Zhou, Z.-M.; Xu, Z.-H. Nitrilases in nitrile biocatalysis: Recent progress and forthcoming research. *Microb. Cell Fact.* **2012**, *11*, 142. [CrossRef] [PubMed]
3. Martínková, L.; Rucká, L.; Nešvera, J.; Pátek, M. Recent advances and challenges in the heterologous production of microbial nitrilases for biocatalytic applications. *World J. Microbiol. Biotechnol.* **2017**, *33*, 8. [CrossRef] [PubMed]
4. Bhalla, T.C.; Kumar, V.; Kumar, V.; Thakur, N.; Savitri. Nitrile metabolizing enzymes in biocatalysis and biotransformation. *Appl. Biochem. Biotechnol.* **2018**, *185*, 925–946. [CrossRef]
5. Martínková, L.; Veselá, A.B.; Rinágelová, A.; Chmátal, M. Cyanide hydratases and cyanide dihydratases: Emerging tools in the biodegradation and biodetection of cyanide. *Appl. Microbiol. Biotechnol.* **2015**, *99*, 8875–8882. [CrossRef]
6. Howden, A.J.M.; Preston, G.M. Nitrilase enzymes and their role in plant-microbe interactions. *Microb. Biotechnol.* **2009**, *2*, 441–451. [CrossRef]
7. Thuku, R.N.; Brady, D.; Benedik, M.J.; Sewell, B.T. Microbial nitrilases: Versatile, spiral forming, industrial enzymes. *J. Appl. Microbiol.* **2009**, *106*, 703–727. [CrossRef]
8. Thimann, K.V., Mahadevan, S.; Nitrilase, I. Occurrence, preparation and general properties of enzyme. *Arch. Biochem. Biophys.* **1964**, *105*, 133–141. [CrossRef]
9. Harper, D.B. Fungal degradation of aromatic nitriles. Enzymology of C–N cleavage by *Fusarium solani*. *Biochem. J.* **1977**, *167*, 685–692. [CrossRef]
10. Goldlust, A.; Bohak, Z. Induction, purification, and characterization of the nitrilase *Fusarium oxysporum* f. sp. *melonis*. *Biotechnol. Appl. Biochem.* **1989**, *11*, 581–601.
11. Brunner, S.; Eppinger, E.; Fischer, S.; Gröning, J.; Stolz, A. Conversion of aliphatic nitriles by the arylacetonitrilase from *Pseudomonas fluorescens* EBC191. *World J. Microbiol. Biotechnol.* **2018**, *34*, 91. [CrossRef] [PubMed]
12. Vorwerk, S.; Biernacki, S.; Hillebrand, H.; Janzik, I.; Müller, A.; Weiler, E.W.; Piotrowski, M. Enzymatic characterization of the recombinant *Arabidopsis thaliana* nitrilase subfamily encoded by the NIT2/NIT1/NIT3-gene cluster. *Planta* **2001**, *212*, 508–516. [CrossRef] [PubMed]
13. Piotrowski, M.; Schönfelder, S.; Weiler, E.W. The *Arabidopsis thaliana* isogene NIT4 and its orthologs in tobacco encode β-cyano-L-alanine hydratase/nitrilase. *J. Biol. Chem.* **2001**, *276*, 2616–2621. [CrossRef] [PubMed]
14. Gong, J.-S.; Li, H.; Zhu, X.-Y.; Lu, Z.-M.; Wu, Y.; Shi, J.-S.; Xu, Z.-H. Fungal His-tagged nitrilase from *Gibberella intermedia*: Gene cloning, heterologous expression and biochemical properties. *PLoS ONE* **2012**, *7*, e50622. [CrossRef]
15. Kaplan, O.; Veselá, A.B.; Petříčková, A.; Pasquarelli, F.; Pičmanová, M.; Rinágelová, A.; Bhalla, T.C.; Pátek, M., Martínková, L. A comparative study of nitrilases identified by genome mining. *Mol. Biotechnol.* **2013**, *54*, 996–1003. [CrossRef]
16. Veselá, A.B.; Rucká, L.; Kaplan, O.; Pelantová, H.; Nešvera, J.; Pátek, M.; Martínková, L. Bringing nitrilase sequences from databases to life: The search for novel substrate specificities with a focus on dinitriles. *Appl. Microbiol. Biotechnol.* **2016**, *100*, 2193–2202. [CrossRef]
17. Shoresh, M.; Harman, G.E.; Mastouri, F. Induced systemic resistance and plant responses to fungal biocontrol agents. *Ann. Rev. Phytopathol.* **2010**, *48*, 21–43. [CrossRef]
18. Kameshwar, A.K.S.; Qin, W. Comparative study of genome-wide plant biomass-degrading CAZymes in white rot, brown rot and soft rot fungi. *Mycology* **2018**, *9*, 93–105. [CrossRef]
19. Hiscox, J.; O´Leary, J.; Boddy, L. Fungus wars: Basidiomycete battles in wood decay. *Stud. Mycol.* **2018**, *89*, 117–124. [CrossRef]

20. Legras, J.L.; Chuzel, G.; Arnaud, A.; Galzy, P. Natural nitriles and their metabolism. *World J. Microbiol. Biotechnol.* **1990**, *6*, 83–108. [CrossRef]
21. Strobel, G.A. The fixation of hydrocyanic acid by a psychrophilic basidiomycete. *J. Biol. Chem.* **1966**, *241*, 2618–2621.
22. Strobel, G.A. 4-Amino-4-cyanobutyric acid as an intermediate in glutamate biosynthesis. *J. Biol. Chem.* **1967**, *242*, 3265–3269.
23. Cabuk, A.; Unal, A.T.; Kolankaya, N. Biodegradation of cyanide by a white rot fungus, *Trametes versicolor*. *Biotechnol. Lett.* **2006**, *28*, 1313–1317. [CrossRef]
24. Luo, H.; Fan, L.; Chang, Y.; Ma, J.; Yu, H.; Shen, Z. Gene cloning, overexpression, and characterization of the nitrilase from *Rhodococcus rhodochrous* tg1-A6 in *E. coli*. *Appl. Biochem. Biotechnol.* **2010**, *160*, 393–400. [CrossRef]
25. Rinágelová, A.; Kaplan, O.; Veselá, A.B.; Chmátal, M.; Křenková, A.; Plíhal, O.; Pasquarelli, F.; Cantarella, M.; Martínková, L. Cyanide hydratase from *Aspergillus niger* K10: Overproduction in *Escherichia coli*, purification, characterization and use in continuous cyanide degradation. *Proc. Biochem.* **2014**, *49*, 445–450. [CrossRef]
26. Felsenstein, J. Confidence limits on phylogenies: An approach using the bootstrap. *Evolution* **1985**, *39*, 783–791. [CrossRef]
27. Nei, M.; Kumar, S. *Molecular Evolution and Phylogenetics*; Oxford University Press: New York, NY, USA, 2000; p. 126.
28. Zhang, L.; Yin, B.; Wang, C.; Jiang, S.; Wang, H.; Yuan, Y.A.; Wei, D. Structural insights into enzymatic activity and substrate specificity determination by a single amino acid in nitrilase from *Synechocystis* sp. PCC6803. *J. Struct. Biol.* **2014**, *188*, 93–101. [CrossRef]
29. Raczynska, J.; Vorgias, C.; Antranikian, G.; Rypniewski, W. Crystallographic analysis of a thermoactive nitrilase. *J. Struct. Biol.* **2010**, *173*, 294–302. [CrossRef]
30. Liu, H.; Gao, Y.; Zhang, M.; Qiu, X.; Cooper, A.J.; Niu, L.; Teng, M. Structures of enzyme-intermediate complexes of yeast Nit2: Insights into its catalytic mechanism and different substrate specificity compared with mammalian Nit2. *Acta Crystallogr. D* **2013**, *69*, 1470–1481. [CrossRef]
31. Thuku, R.N.; Weber, B.W.; Varsani, A.; Sewell, B.T. Post-translational cleavage of recombinantly expressed nitrilase from *Rhodococcus rhodochrous* J1 yields a stable, active helical form. *FEBS J.* **2007**, *274*, 2099–2108. [CrossRef]
32. Fernandes, B.C.M.; Mateo, C.; Kiziak, C.; Chmura, A.; Wacker, J.; van Rantwijk, F.; Stolz, A.; Sheldon, R.A. Nitrile hydratase activity of a recombinant nitrilase. *Adv. Synth. Catal.* **2006**, *348*, 2597–2603. [CrossRef]
33. Yeom, S.J.; Kim, H.J.; Lee, J.K.; Kim, D.E.; Oh, D.K. An amino acid at position 142 in nitrilase from *Rhodococcus rhodochrous* ATCC 33278 determines the substrate specificity for aliphatic and aromatic nitriles. *Biochem. J.* **2008**, *415*, 401–407. [CrossRef]
34. Martínková, L. Nitrile metabolism in fungi: A review of its key enzymes nitrilases with focus on their biotechnological impact. *Fung. Biol. Rev.* **2019**, *33*, 149–157. [CrossRef]
35. Hibbett, D.S. Agaricomycotina. Jelly Fungi, Yeasts, and Mushrooms. Version 20 April 2007. Available online: http://tolweb.org/Agaricomycotina/20531/2007.04.20 (accessed on 22 October 2019).
36. Ghosh, S.; Gupta, S.K.; Jha, G. Identification and functional analysis of AG1-IA specific genes of *Rhizoctonia solani*. *Curr. Genet.* **2014**, *60*, 327–341. [CrossRef]
37. Abdullah, A.S.; Moffat, C.S.; Lopez-Ruiz, F.J.; Gibberd, M.R.; Hamblin, J.; Zerihun, A. Host-multi-pathogen warfare: Pathogen interactions in co-infected plants. *Front. Plant Sci.* **2017**, *8*, 1806. [CrossRef]
38. Wang, P.; Van Etten, H.D. Cloning and properties of a cyanide hydratase gene from the phytopathogenic fungus *Gloeocercospora sorghi*. *Biochem. Biophys. Res. Commun.* **1992**, *187*, 1048–1054. [CrossRef]
39. Basile, L.J.; Willson, R.C.; Sewell, B.T.; Benedik, M.J. Genome mining of cyanide degrading nitrilases from filamentous fungi. *Appl. Microbiol. Biotechnol.* **2008**, *80*, 427–435. [CrossRef]
40. Edgar, R.C. MUSCLE: Multiple sequence alignment with high accuracy and high throughput. *Nucleic Acids Res.* **2004**, *32*, 1792–1797. [CrossRef]
41. Kumar, S.; Stecher, G.; Li, M.; Knyaz, C.; Tamura, K. MEGA X: Molecular evolutionary genetics analysis across computing platforms. *Mol. Biol. Evol.* **2018**, *35*, 1547–1549. [CrossRef]
42. Altschul, S.F.; Gish, W.; Miller, W.; Myers, E.W.; Lipman, D.J. Basic local alignment search tool. *J. Mol. Biol.* **1990**, *215*, 403–410. [CrossRef]

43. Notredame, C.; Higgins, D.G.; Heringa, J. T-Coffee: A novel method for fast and accurate multiple sequence alignment. *J. Mol. Biol.* **2000**, *302*, 205–217. [CrossRef] [PubMed]
44. Clamp, M.; Cuff, J.; Searle, S.M.; Barton, G.J. The Jalview Java alignment editor. *Bioinformatics* **2004**, *20*, 426–427. [CrossRef] [PubMed]
45. Krieger, E.; Koraimann, G.; Vriend, G. Increasing the precision of comparative models with YASARA NOVA a self-parameterizing force field. *Proteins* **2002**, *47*, 393–402. [CrossRef] [PubMed]
46. Sali, A.; Blundell, T.L. Comparative protein modelling by satisfaction of spatial restraints. *J. Mol. Biol.* **1993**, *234*, 779–815. [CrossRef]
47. Davis, I.W.; Leaver-Fay, A.; Chen, V.B.; Block, J.N.; Kapral, G.J.; Wang, X.; Murray, L.W.; Arendall, W.B., III; Snoeyink, J.; Richardson, J.S.; et al. MolProbity: Allatom contacts and structure validation for proteins and nucleic acids. *Nucleic Acids Res.* **2007**, *35*, 375–383. [CrossRef]
48. *Schrödinger Release 2017-1: Canvas*; Schrödinger LLC: New York, NY, USA, 2017.
49. Cluness, M.J.; Turner, P.D.; Clements, E.; Brown, D.T.; O'Reilly, C. Purification and properties of cyanide hydratase from *Fusarium lateritium* and analysis of the corresponding chy1 gene. *J. Gen. Biol.* **1993**, *139*, 1807–1815. [CrossRef]
50. Black, G.W.; Brown, N.L.; Perry, J.J.B.; Randall, D.; Turnbull, G.; Zhang, M. A high-throughput screening method for determining the substrate scope of nitrilases. *Chem. Commun.* **2015**, *51*, 2660–2662. [CrossRef]
51. Olsen, B.A. Hydrophilic interaction chromatography using amino and silica columns for the determination of polar pharmaceuticals and impurities. *J. Chromatogr. A* **2001**, *913*, 113–122. [CrossRef]

MDPI
St. Alban-Anlage 66
4052 Basel
Switzerland
Tel. +41 61 683 77 34
Fax +41 61 302 89 18
www.mdpi.com

International Journal of Molecular Sciences Editorial Office
E-mail: ijms@mdpi.com
www.mdpi.com/journal/ijms